Modern Civil Engineering Structures

Dr. D.K. Chauhan

RANDOM
RANDOM PUBLICATIONS
NEW DELHI (INDIA)

Modern Civil Engineering Structures

ISBN 978-93-5111-568-7

Published in 2015 in India by

RANDOM PUBLICATIONS

4376-A/4B, Gali Murari Lal, Ansari Road
New Delhi-110 002
Phone : +9111-43580356, 011-23289044, 011-43142548
e-mail: sales@randompublications.com,
info@randompublications.com, randomexports@gmail.com

Reprinted 2022

Type Setting by : Friends Media, Delhi-110089
Digitally Printed at: Replika Press Pvt. Ltd.

Preface

A large part of civil engineering deals with the structures which define our society. This has been true throughout the history of mankind. Today, the structures that ancient engineers have left behind help us understand the people who lived before us, how they lived, the wars that they fought and the gods that they served.

Civil Engineering is one of the broadest and oldest of the engineering disciplines, extending across many technical specialties. Civil Engineers plan, design, and supervise the construction of facilities essential to modern life like space satellites and launching facilities, offshore structures, bridges, buildings, tunnels, highways, transit systems, dams, airports, harbors, water supply system and wastewater treatment plants. A civil engineer is responsible for planning and designing a project, constructing the project to the required scale, and maintenance of the project.

In past years, civil engineers focused on design and construction of new facilities, such as buildings, bridges and highways, water treatment and environmental facilities, foundations and tunnels. Today's civil engineer not only has to design new facilities but must also analyze the effects of deterioration on infrastructure elements, consider system interdependencies and evaluate life-cycle impacts while also considering environmental and economic sustainability within the context of society. Civil engineers must be equipped with in-depth knowledge of traditional, fundamental principals and new technologies in order to address the complex, interdisciplinary problems faced within society. The undergraduate program at CUA gives the students the necessary background to success within this new context and to become the future leaders of the profession.

Civil engineering has a significant role in the life of every human being, though one may not truly sense its importance in our daily routine. The function of civil engineering commences with the start of the day when we take a shower, since the water is delivered through a water supply system including a well designed network of pipes, water treatment plant and other numerous associated services. The network of roads on which we drive while proceeding to school or work, the huge structural bridges we come across and the tall buildings where we work, all have been designed and constructed by civil engineers. Even the benefits of electricity we use are available to us through the contribution of civil engineers who constructed the towers for the transmission lines. In fact, no sphere of life may be identified that does not include the contribution of civil engineering. Thus, the importance of civil engineering may be determined according to its usefulness in our daily life.

Civil Engineering covers all aspects of civil engineering and its role in society. This book is integrates all of the diverse approaches that have gone into the making of the modern field of civil engineering.

I would like to thank my team for standing beside me throughout my career and writing this book. My special thanks go to "Random Publications" who have published the book.

– Dr. D.K. Chauhan

Contents

1

Introduction to Civil Engineering

Civil engineers design and build many of the things that we use every day. Civil engineers work in the following fields: structures, transport, energy and water. Many civil engineers work in the building industry together with architects, surveyors and building services engineers when designing structures and foundations for large buildings.

Civil engineers design transport systems: they build roads, stations and airports. They also design systems for controlling traffic to ensure effective and safe use of the road system and aim to reduce the impact of roads and traffic on the environment. Civil engineers design and construct all types of power stations and hydroelectric projects. Civil engineers are also involved in the research that is taking place into the harnessing of renewable sources of energy such as geothermal and wave power. Civil engineers design dams to create water storage reservoirs. Hydropower is used to generate electricity or the water may be diverted through tunnels, canals and pipelines to provide irrigation for the land, to transport water to cities, supply safe drinking water to home and industry after it has been treated. Engineers also provide the systems for the collection and treatment of sewage so that it can be returned to the environment without causing pollution.

HISTORY OF CIVIL ENGINEERING

The history of civil engineering can be traced back to 4000 BC when the sole means of construction was human Labour, lacking any sophisticated equipment. With advancement in all spheres of technology, civil engineering has also developed tremendously.

MANUAL LABOUR: THE FIRST ENGINEERING TOOL

Civil engineering involves the design, construction, and maintenance of works such as roads, bridges, and buildings. It's a science that includes a variety of disciplines including soils, structures, geology, and other fields. Thus the history of civil engineering is closely associated with the history of advancement in these sciences. In ancient history, most of the construction was carried out by artisans, and technical expertise was limited. Tasks were accomplished by the utilization of manual Labour only, without the use of sophisticated machinery, since it did not exist. Therefore, civil engineering works could only be realized with the utilization of a large number of skilled workers over an extended period of time.

PREHISTORIC AND ANCIENT CIVIL ENGINEERING STRUCTURES

It might be appropriate to assume that the science of civil engineering truly commenced between 4000 and 2000 BC in Egypt when transportation gained such importance that it led to the development of the wheel. According to the historians, the Pyramids were constructed in Egypt during 2800-2400 BC and may be considered as the first large structure construction ever. The Great Wall of China that was constructed around 200 BC is considered another achievement of ancient civil engineering. The Romans developed extensive structures in their empire, including aqueducts, bridges, and dams. A scientific approach to the physical sciences concerning civil engineering was implemented by Archimedes in the third century BC, by utilizing the Archimedes Principle concerning buoyancy and the Archimedes screw for raising water.

THE ROLES OF CIVIL AND MILITARY ENGINEER IN ANCIENT TIMES

As stated above, civil engineering is considered to be the first main discipline of engineering, and the engineers were in fact military engineers with expertise in military and civil works. During the era of battles or operations, the engineers were engaged to assist the soldiers fighting in the battlefield by making catapults, towers, and other instruments used for fighting the enemy. However, during peace time, they were concerned mainly with the civil activities such as building fortifications for Defence, making bridges, canals, etc.

CIVIL ENGINEERING IN THE 18TH - 20TH CENTURY

Until the recent era, there was no major difference between the terms civil engineering and architecture, and they were often used interchangeably. It was in the 18th century that the term civil engineering was firstly used independently from the term military engineering. The first private college in the United States that included Civil Engineering as a separate discipline was Norwich University established in the year 1819. Civil engineering societies were formed in United States and European countries during the 19th century, and similar institutions were established in other countries of the world during the 20th century. The American Society of Civil Engineers is the first national engineering society in the United States. In was founded in 1852 with members related to the civil engineering profession located globally. The number of universities in the world that include civil engineering as a discipline have increased tremendously during the 19th and the 20th centuries, indicating the importance of this technology.

MODERN CONCEPTS IN CIVIL ENGINEERING

Numerous technologies have assisted in the advancement of civil engineering in the modern world, including high-tech machinery, selection of materials, test equipment, and other sciences. However, the most prominent contributor in this field is considered to be computer-aided design (CAD) and computer-aided manufacture (CAM). Civil engineers use this technology to achieve an efficient system of construction, including manufacture, fabrication, and erection. Three-dimensional design software is an essential tool for the civil engineer that facilitates him in the efficient designing of bridges, tall buildings, and other huge complicated structures.

FIELD OF CIVIL ENGINEERING

Civil engineering is the oldest and one of the most versatile branches of engineering. Every structure that we see around today is a creation of civil engineering. Learn more about this fascinating discipline.

Civil engineering is the oldest and broadest engineering discipline among all the engineering fields. The field deals with the planning, designing, and construction of buildings and various other structures. From huge dams to sky high buildings, from suspension bridges to offshore drilling platforms, every physical concrete structure comes under civil engineering.

The usage of civil engineering dates back to the ancient times. Most of the seven wonders, including Egyptian pyramids and the Taj Mahal, are the creation of offshore civil engineering skills. Some of the world's oldest civilizations such as Harappa and Mohenjo-daro are also the product of immaculate civil engineering. After all, how can one ignore the magnificent architecture of Rome's Coliseum or the unique design of the great Eiffel tower?

SUB-BRANCHES OF CIVIL ENGINEERING

Civil engineering is the most diverse field of all the engineering branches. Technically speaking, civil engineering is the creator of the whole world's infrastructure. Structures such as tunnels, dams, sewers, bridges, highways, canals, industrial plants, residential buildings, railway lines, airports etc come under the category of civil engineering.

Moreover, as the population of the world increases and the technology becomes more advanced, the need for better infrastructure will increase around the world. Civil engineering continues to cater to these needs in all the sectors and aspects of human life.

The need for infrastructure has increased by leaps and bounds in each and every sector. In order to concentrate and manage the construction process in each sector, the field of civil engineering has been divided into various sub-disciplines. This means that on the basis of applications, the stream of civil engineering has been segregated into several branches to make the construction process easier and more manageable. Some of the main branches include – bridge engineering, construction engineering, coastal engineering, geotechnical engineering, environmental engineering, transportation engineering, surveying etc.

APPLICATION AND PROCESS

An ideal process of civil engineering will include the construction of a concrete structure right from the planning to the maintenance of the structure even after it is made. Civil engineers are the professionals who perform the functions involved in a construction process.

They plan, design and analyze each and every part of the structure before starting the actual construction process. Designs of all the parts of the building along with precise specifications are drafted on drawing sheets prior to the construction process. A study of various forces that can act at each of the building corners is made and on the basis of that further modification in the design is done. Even the minutest corner of the planned structures is scrutinized to check the amount of stresses and strains that might arise when the structure is built or in the process of making. Apart from this, the structures are also provided with tolerance levels to adjust deviations due to wind, seismic activities, or natural calamities. These tolerances allow the building to resist the natural forces and thus stand strong without being impacted.

Thus civil engineering deals with setting up reliable structures with guaranteed longevity. Also, with an increase in awareness towards renewable sources of energy, the applications of civil engineering have also increased. Environmental engineering, a sub-branch of civil engineer, deals with constructing structures that have low impact on the environment and leave no carbon foot prints.

Environmental engineering also deals with creating structures on both onshore and of shore sides in order to make efficient and environmentally friendly sources of energy.

Some of these applications include various methods to purify the contaminated air and water, utilizing solar energy, generating fresh water, utilizing wind and water energy, and protecting the marine environment.

Civil engineering is becoming more and more diversified with the increase of applications. Moreover, the guidelines of constructing structures have also become more stringent, stressing issues such as human safety and resistance to natural and man-made calamities.

CONCEPT OF CIVIL ENGINEERING

In the starting of era, Civil Engineering accomodates all engineers that did not practice military engineering; but said to have begun in 18th century in France.

- First Civil Engineer was an Englishman, named John Smeaton in the year 1761
- Civil engineers recorded and saved more families and lives as in comparison to the doctors in history development of clean water and sanitation systems
- First KY Civil Engineering Graduate Henry H. White, from Bacon (Georgetown) finished his degree in the year 1840
- In the beginning during the year 1886, State College established civil engineering degree.
- Er.John Wesley Gunn of Lexington received first Civil Engineering degree from A and M College in the year 1890

It was believed that, engineering has been an aspect of life since the beginnings of human existence.

The basic practice of civil engineering may have commenced between the era 4000 and 2000 BC in Ancient Egypt and Mesopotamia, when humans started to dump a nomadic existence, by creating a need for the construction of shelter and roof. At the same time, transportation became increasingly important which has shown the way to the development of wheel and sailing. In anticipation of modern times, there was no quick and right distinction among the civil engineering and architecture, and the result was that the term engineer and architect were mainly referring to the geographical variations being the same person, often used interchangeably.

At the time of construction of Pyramids in Egypt, was the first instances of large structure constructions. Other ancient historic civil engineering constructions includes:

- Qanat water management system which was considered as the oldest and the longest
- Parthenon by Iktinos in Ancient Greece
- Appian Way by Roman engineers
- Great Wall of China by General Meng T'ien
- stupas of ancient Sri Lanka
- extensive irrigation works in Anuradhapura.

It was found that the Romans developed civil structures throughout their empire, which includes especially

- Aqueducts
- Insulae
- Harbors
- Bridges
- Dams
- Roads

As considered, civil engineering is one of the oldest among all engineering professions. It was believed that the ancient feats such as the construction of the Egyptian pyramids and roman road network systems are basically based on the principles that involved civil construction and maintenance.

Civil engineers as the backbone of the engineering can be found in all areas of society namely:

- private contractors
- municipal corporations
- federal government organizations
- military.

CIVIL ENGINEERING FOCUS:

Civil engineering is believed to be much broader and wider field that has a number of specialties which covers:

- Structural
- Construction
- Environmental
- Hydraulics
- Transportation.

Civil engineering mainly focuses on the design, construction and maintenance of everything from buildings and roads to pipelines and power plants. Civil engineers plan and design various infrastructure that is required or needed to develop and maintain facilities, services and organizational structures for cities and communities.

Civil engineers constructed the world's infrastructure. For this, they quietly shape the history of nations around the world. Many people can not even imagine life without the varied contributions done by civil engineers to the public's health, safety and standard of living.

Thereby by exploring civil engineering's influence in shaping the world in todays time, that led to creatively imagine the growth and progress of our tomorrows.

It was seen that the civil engineering mainly focuses on the infrastructure of the world in the areas related to:

- Water works
- Sewers
- Dams
- Power Plants
- Transmission Towers
- Railroads
- Highways
- Bridges
- Tunnels
- Irrigation Canals
- River Navigation
- Shipping Canals
- Traffic Control
- Mass Transit
- Airport Runways
- Terminals
- Industrial Plant Buildings
- Skyscrapers

PROCESSES OF CIVIL ENGINEERING

We believed that the planning, designing, and construction of facilities that serve up people are nothing but civil engineering. Such type of facilities include the highways that are nowadays used to connect our nation's cities, airports that serve our travelers, bridges that spreads our rivers and harbors, dams and levees which helps in controlling floods and supply water to cities, and sewage treatment plants that helps us to protect our environment. By comparing the working with architects and engineers from other engineering, the civil engineer also participates in the designing and construction of buildings and structures.

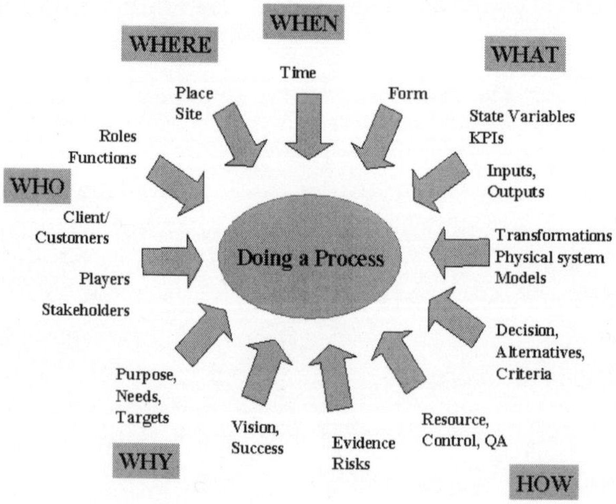

Initially the civil engineering process begins with the accumulation and analysis of basic information and data. These data include the information related to:

- topography and geology in case of highway route
- flood history of a river that must be bridged or dammed
- information on population growth
- earthquake history
- analysis of construction materials
- pollution and surveys related to air, land, and water

The collection and analysis of these kinds of data are totally required for any type of planning, whether it is for future needs or is needed for a particular project. With these data, civil engineers design and use their knowledge of science and engineering by drafting a layout plan to meet a project's requirements to make it a successful completion.

Civil Engineering basically follow process such as:

- Planning ...
- Design ...
- Construction ...
- Operation/Maintenance ...
- Rehabilitation ...

CIVIL ENGINEERING SPECIALTIES

Civil engineering covers several specialties that are viewed and analysed everywhere.

The broader civil engineering education covers knowledge in varied specialties. General civil engineers are usually responsible for designing and layouting of plan that includes:

- Grading
- utility design

- environmental compliance
- project management

It was observed that the general civil engineers also become registered professional land surveyors and are engaged in various surveying and civil activitries. The various specific area of civil engineering domains are:

- Structural
- Surveying
- Transportation
- Water Resources
- Construction
- Environmental/ Water Quality
- Geotechnical
- Hydraulics
- Materials

SCOPE OF CIVIL ENGINEERING

Civil engineering being the oldest and the reputed branch of engineering that has gained importance after military period. In our life, many things are related directly or indirectly to the product of civil engineering.

It is a field of engineering sciences that is related to:

- Design
- Construction
- Maintenance of buildings, dams, bridges, tunnels, highways and various related structures by the use of engineering laws, mathematical evaluations and theory of mechanics. Civil Engineers uses the required available resources such as expertise, materials, manpower in order to finish the project with accuracy within a given period of time span by keeping in view the quality, cost, physical and natural issues and chemical and biochemical hazards of the specific project.

Inspite of increase in the scope and importance of civil engineering, it has now got diversified into many branches of skills and study. Some of the significant study areas are:

- structural engineering
- geotechnical engineering
- transportation engineering
- hydraulic engineering
- environmental engineering

For a particular project, engineers are appointed by many companies and by contractors that used to take small or large engineering contract or even can go up for businesses tieups on a new invention idea to large-scale that work on immense contracts.

FACTORS USED IN CONSTRUCTION

It was seen that there arises five factors that are required to be taken into consideration while designing or evaluating a building as it applies to how it will survive fire and fire extension.

Type of Construction

By understanding the type of construction, we can judge the quality and the time required to execute the work.

Size of the Building

The size of the building will allow the type of load that the building will support. Accordingly the structure specification has to be designed. The physical dimensions of the building, and knowledge of the location will allow incoming companies to estimate the quality and mixer used.

Age of Construction

The age of construction will help measure whether it is conventional or lightweight construction. Lightweight construction tends to fail much more quickly. By knowing the age of the structure or the building will help to evaluate features that are specific to the type of construction. As the life of a building is between 75 to 100 years, the erosion of mortar in brick walls, corrosion of exposed metals, rotting or insect infestation of wooden structural members all will contribute to degradation of the structural integrity.

Renovations

In case of renovations of a building, it is taken into account that a specific type of floor loading is required. The material used for structure should be mixed and mortar to be added in proportion in order to give hardness. Remodeling a building makes determination of age and construction more difficult, and requires even more diligence during inspections, to get an accurate handle on the true type of construction present.

Type of Occupancy

The type of occupancy allows a determination of what kind and number will be there to support the actual load structure.

CIVIL ENGINEERING BENEFITS

Civil engineers basically help in planning along with construction of major structures, such as:

- Subways
- Bridges
- Airports

- Highways
- Tunnels
- Buildings

By safely completion of the particular projects is usually required and extensive analysis has to be performed in order to identify the most useful and accurate materials for construction.

Identification

Junior or Site Engineers mostly work along with the senior architects on construction site and on particular design, and further supervise the construction working alongside. Mostly it is under the power of the civil engineers to decide on:

- appropriate methods and materials
- manage budgets
- review project drawings
- schedule materials
- equipment purchases

Building inspection

It is required that specialization is of ultimate importance in areas related to:

- Construction
- Structural
- Environmental

For example, hydraulic engineers design dams, canals, harbors, flood control systems, reservoirs and irrigation methods. Similarly building inspection and extensive research, civil engineers can develop new water sources. In case of Transportation engineers, supervise the construction of tunnels, bridges, highways, and design longer-lasting constructions.

Better employment opportunities

Basically good and intelligent, successful civil engineers work for the government. In developing and developed nation, civil activities will continue, so the demand of civil engineers was always high. More importantly, engineers design storage and utility companies.

Higher remuneration

Civil engineers gets higher paid up normally around Rs 5,00,000 per year, as per the statistics done. While it takes more than five years of working experience to become a professional civil engineer. The starting salaries typically range between Rs 25,000 to 35,000. Moreover, advancement in civil engineering translates to better pay and supervisory roles.

The benefits of civil engineering can be more commonly explained as:

- Solve societal problems
- Allows indoor and outdoor work
- Responsible, highly respected job
- Challenging technical career
- Utilize modern technology
- Work with people of various backgrounds
- Well paid

CIVIL ENGINEERING STRUCTURES

A large part of civil engineering deals with the structures which define our society. This has been true throughout the history of mankind. Today, the structures that ancient engineers have left behind help us understand the people who lived before us, how they lived, the wars that they fought and the gods that they served.

In a sense, we can say that the earliest civil engineers were the architects who designed Egypt's pyramids and their contemporaries in other lands. Although the structures which they engineered in those days are extremely inefficient in their use of materials by today's standards, they are still engineering marvels.

Rome's aqueducts and roads were an engineering marvel in their day. To be able to transport water over such long distances, without the use of pumps and pipes was a true technological challenge. By understanding the physics of liquid flow, and combining it with the structural stability of the arch, those engineers were able to provide water to areas that otherwise couldn't be farmed.

Even today, the work of civil engineers, modifies the environment around us, creates space where we can live, work and play. While the structures which we see most commonly are buildings, civil engineering deals with many other structures as well, include dams, roads, bridges, tunnels, water treatment plants and more.

DAMS

The Hoover dam, the second largest dam in the United States was an engineering marvel in its day. It is sited on the Colorado River, at the border between Arizona and Nevada, creating Lake Mead, which stretches 110 miles upstream to the Grand Canyon. This dam takes advantage of the naturally existing canyon, which is an extension of the Grand Canyon to provide its foundation as well as room for the lake it created. Upon its completion in 1938, it was the tallest dam in the world. Although many have since surpassed it, it is still in the top 30.

One of the great challenges which faced the civil engineers in the creation of the Hoover Dam was in diverting the flow of the Colorado River while the dam was being created. To accomplish this, they dug four diverter tunnels through the solid rock of the canyon walls. Once completed, a cofferdam was erected upstream of the dam's location, to divert the water through these tunnels.

At the time of its building, the Hoover Dam required more concrete than had ever been used in a single structure. Engineers on the project calculated that such a large volume of concrete would require 75 years to fully set. Since that was unacceptable, they designed a series of cooling pipes, carrying cold water throughout the dam, cooling the concrete. This allowed it to set considerably faster. At the time, this was a revolutionary method, which had never been attempted before and was invented just for the construction of the Hoover Dam.

BUILDINGS

The world's tallest building, the Burj Khalifa off the coast of Dubai in the United Arab Emirates is both the tallest building and the tallest man made structure in the world. At 2,723 feet tall, it created special challenges for the civil engineering team who designed it. Amongst other problems to confront, was that they were building the world's tallest building on sand. Added to this were the high winds which upper floors of the building would be subjected to. The secret to the Burj Khalifa's success as a structure is a combination of its foundation and its core. The foundation consists of 192 holes, bored 1.5 meters in diameter and 50 meters deep. These pilings are all connected to a 12 foot thick raft-type foundation, which is as big as the entire footprint of the building.

Besides the foundation, the other major structural element of any skyscraper is the core. This reinforced cement structural element is in the center of the floor, and typically contains the elevator shafts and stairwells. However, its primary purpose is to support the floors and skin of the building. Burj Khalif's "Y" shape is intended to maximize the strength of the core, by using a hexagonal core, combined with columns at the exterior points of the "Y." These are connected together by outrigger walls on the mechanical floors, providing the same effect as buttresses on Medieval Cathedrals and castles. This combination provides excellent strength for both vertical and lateral loads.

The building is also designed with a slight twist in the "Y" as it goes up, which is designed to help combat lateral wind forces. By adding a twist, the structural engineers were able to ensure that wind would not be able to strike one face of the building throughout its entire height. That reduces the potential lateral forces on the structure. All in all, the Burj Khalif is a remarkable feat of structural engineering.

BRIDGES

Building bridges creates a totally different type of challenge to the civil engineer. In the case of a suspension or cable-stayed bridge, a large part of the completed bridge's load is carried by the cables. However, those cables can't properly support the bridge while it is under construction. Throughout the course of its construction, the sections of the bridge are cantilevered, solely supported by their own structural strength and attachment to the towers.

Successful suspension bridge designs combine various structural elements to carry their load. The purpose of the suspension cables is to transfer the weight of the bridge deck or roadbed and vehicle load to the towers, which are anchored to bedrock. Since both steel and concrete are extremely strong under compression, this allows even delicate looking towers to carry an incredible amount of load.

For the suspension cables to be able to transfer the bridge's load to the towers, the cables need to be under tension. However, without a complete roadbed, the cables cannot be taut. This means that the roadbed must support itself, until it is joined together in the midst of the span and all the cables are in place.

To accommodate this need and provide stiffness to the roadbed, trusses are attached to the underside of the roadbed. Without sufficient truss strength to stiffen the roadbed, it would undulate like the Tacoma Narrows Bridge, which collapsed in 1940.

Construction of these bridges has to be accomplished by working from the towers outward. The distance between the towers and their distance from the bridge's ends is very carefully calculated by the structural engineers, so that the roadbed can be built in both directions at the same time. To try and build the bridge only in one direction from the tower, without building it in the other direction, would cause the structure to become out of balance, with the towers leaning towards the side that has the weight of the roadbed attached to it. This would cause permanent damage to the columns.

Each of these structures, like every other structure which civil engineers work on, has its own design challenges. The civil engineers who work on these projects start by understanding the problems which they are faced with. The design process consists of finding solutions to those problems, then challenging those solutions, to ensure that the resulting structures will be safe.

SCOPE AND DIFFERENT FIELD OF CIVIL ENGINEERING

Outstanding to increase in the scope of civil engineering with the passageway of time, it accepts now acquired diversified into many branches of learning. Approximately of the significant ones include structural engineering, Geotechnical engineering, transportation engineering, hydraulic engineering, environmental engineering and a few more significant areas of analysis.

Engineers are engaged by a wide brows of companies in the United States, from small start up businesses focused on a new invention estimate to large-scale companies that work on immense contracts. Engineers from different fields constantly work together to create successful products. Once considering the design and manufacture of an aircraft, for example, the workforce behind the evolution will include aeronautical engineers optimizing airflow paths, analysis engineers evaluating the strength of landing gear developed by design engineers, electronics engineers developing wiring methods and pilot ascendances, ergonomic engineers designing comfortable seating and computer engineers programming the aircraft operation systems, including everything from the autopilot system to the cabin crew call system.

Apart from structures on land and general transportation systems, civil engineers are also responsible for building good transportation systems for flow of water, *i.e.* the water distribution systems. The main activities in this undertaking are designing pipelines for flow of water, drainage facilities, canals, dams, etc. Dams are a major source for non-conventional source of electricity and are hence in high demand today. While designing these structures, the civil engineers take into account the various properties of fluids to calculate the forces acting at different points.

SUB-DISCIPLINES OF CIVIL ENGINEERING

Civil engineering is a multiple science encompassing numerous sub-disciplines that are closely linked with each other.

STRUCTURAL ENGINEERING

This discipline involves the design of structures that should be safe for the users, be economical, and accomplish the desired functions. The design and analysis should initially identify the loads that act on the structures, stresses that are created due to loads, and then design the structure to withstand these loads. It includes steel structures, buildings, tunnels, highways, dams, and bridges.

GEOTECHNICAL ENGINEERING

Geotechnical engineering deals with soils, rocks, foundations of buildings and bridges, highways, sewers and underground water systems. Technical information obtained from the sciences of geology, material testing, and hydraulics is applied in the design of foundations and structures to ensure safety and economy of construction.

WATER RESOURCES ENGINEERING

This discipline of civil engineering concerns the management of quantity and quality of water in the underground and above ground water resources, such as rivers, lakes and streams.

Geographical areas are analyzed to forecast the amount of water that will flow into and out of a water source. Fields of hydrology, geology, and environmental science are included in this discipline of civil engineering.

Environmental Engineering: : It is related to the science of waste management of all types, purification of water, cleaning of contaminated areas, reduction of pollution, and industrial ecology. Technical data obtained due to environmental engineering assists the policy makers in making decisions related to environmental issues.

OTHER DISCIPLINES

Some of the other disciplines included in civil engineering include coastal engineering, construction engineering, earthquake engineering, materials science, transportation engineering, and surveying.

FUTURE OF CIVIL ENGINEERING

Civil engineering utilizes technical information obtained from numerous other sciences, and with the advancement in all types of technologies, the civil engineering has also benefited tremendously. The future of civil engineering is expected to be revolutionized by the new technologies including design software, GPS, GIS systems and other latest technical expertise in varied fields. Technology will continue to make important changes in the application of civil engineering, including the rapid progress in the use of 3-D and 4-D design tools.

2

Architecture in Civil Engineering

Civil engineers and architects complement each other's' work. An architect takes care of the design and shape of a building whereas a civil engineer takes care of the technical nitty-gritty like the strength of the building. Architecture is art; civil engineering is more about physics.

The primary focus of an architect remains on the aesthetic appearance of a building. Architecture is closely related to Civil Engineering, and as recently as a few decades ago, there were no architects as a separate profession at all.

Properly implemented concepts of architecture ensure that a building is not just strong, but that it looks good and takes care of the comfort of the inhabitants as well.

A civil engineer need not bother about the looks, design, in-house lighting, greenery, and other things when an architect is around. That leaves him with the primary job of making a building stable and strong. Other than mastering the concepts of architecture, an architect is supposed to have a basic knowledge of structural engineering as well.

THE WORLD OF ARCHITECTURE: FUNCTIONS OF AN ARCHITECT, SCOPE, AND ADVANCEMENTS

An architect is just not a helping hand to the civil engineer, but much more than that. The planning and designing of any construction project is incomplete without the input of an architect. The establishment of buildings and land with reference to functional and aesthetic requirements is a primary function of an architect. Architectural science has become a vast subject, and today we have architecture styles of different kinds.

Urban architecture, rural architecture, contemporary architecture, and modern architecture are some of the styles that are studied across the globe. The inclusion of software tools like 3D Max, Maya, CAD, and CAM has made things easier for both the civil engineers and architects because graphical representation of any kind of work always adds efficiency and better understanding.

HISTORICAL BACKGROUND OF ARCHITECTURE

Fig. Brunelleschi, in the building of the dome of Florence Cathedral in the early 15th-century, not only transformed the building and the city, but also the role and status of the architect.

Fig. Section of Brunelleschi's dome drawn by the architect Cigoli (c. 1600)

Architecture is both the process and the product of planning, designing, andconstructing buildings and other physical structures. Architectural works, in the material form ofbuildings, are often perceived as cultural symbols and as works of art. Historical civilizations are often identified with their surviving architectural achievements.

"Architecture" can mean:

- A general term to describe buildings and other physical structures.
- The art and science of designing buildings and (some) Non-building structures.
- The style of design and method of construction of buildings and other physical structures.
- The knowledge of art, science and technology and humanity.
- The practice of the architect, where architecture means offering or rendering professional services in connection with the design and construction of buildings, or built environments.

- The design activity of the architect, from the macro-level (urban design, landscape architecture) to the micro-level (construction details and furniture).

Architecture has to do with planning, designing and constructing form, space and ambience to reflect functional, technical, social, environmental and aesthetic considerations. It requires the creative manipulation and coordination of materials and technology, and of light and shadow. Often, conflicting requirements must be resolved. The practise of Architecture also encompasses the pragmatic aspects of realizing buildings and structures, including scheduling, cost estimation and construction administration. Documentation produced by architects, typically drawings, plans and technical specifications, defines the structure and/or Behaviour of a building or other kind of system that is to be or has been constructed.

The word "architecture" has also been adopted to describe other designed systems, especially in information technology.

THEORY OF ARCHITECTURE

HISTORIC TREATISES

The earliest surviving written work on the subject of architecture is *De architectura*, by the Roman architect Vitruvius in the early 1st century AD. According to Vitruvius, a good building should satisfy the three principles of *firmitas, utilitas, venustas*, commonly known by the original translation – *firmness, commodity and delight*. An equivalent in modern English would be:

- Durability – a building should stand up robustly and remain in good condition.
- Utility – it should be suitable for the purposes for which is it used.
- Beauty – it should be aesthetically pleasing.

According to Vitruvius, the architect should strive to fulfill each of these three attributes as well as possible. Leone Battista Alberti, who elaborates on the ideas of Vitruvius in his treatise, De Re Aedificatoria, saw beauty primarily as a matter of proportion, although ornament also played a part. For Alberti, the rules of proportion were those that governed the idealised human figure, the Golden mean. The most important aspect of beauty was therefore an inherent part of an object, rather than something applied superficially; and was based on universal, recognisable truths. The notion of style in the arts was not developed until the 16th century, with the writing of Vasari: by the 18th century, his Lives of the Most Excellent Painters, Sculptors, and Architects had been translated into Italian, French, Spanish and English.

The Houses of Parliament, Westminster, master-planned byCharles Barry, with interiors and details by A.W.N. Pugin

In the early 19th century, Augustus Welby Northmore Pugin wrote *Contrasts* (1836) that, as the titled suggested, contrasted the modern, industrial world, which he disparaged, with an idealized image of neo-medieval world. Gothic architecture, Pugin believed, was the only "true Christian form of architecture."

The 19th-century English art critic, John Ruskin, in his *Seven Lamps of Architecture*, published 1849, was much narrower in his view of what constituted architecture. Architecture was the "art which so disposes and adorns the edifices raised by men... that the sight of them" contributes "to his mental health, power, and pleasure".

For Ruskin, the aesthetic was of overriding significance. His work goes on to state that a building is not truly a work of architecture unless it is in some way "adorned". For Ruskin, a well-constructed, well-proportioned, functional building needed string courses or rustication, at the very least.

On the difference between the ideals of *architecture* and mere *construction*, the renowned 20th-century architect Le Corbusier wrote: "You employ stone, wood, and concrete, and with these materials you build houses and palaces: that is construction. Ingenuity is at work. But suddenly you touch my heart, you do me good. I am happy and I say: This is beautiful. That is Architecture". Le Corbusier's contemporary Ludwig Mies van der Rohe said "Architecture starts when you carefully put two bricks together. There it begins."

The National Congress of Brazil, designed by Oscar Niemeyer.

MODERN CONCEPTS OF ARCHITECTURE

The notable 19th-century architect of skyscrapers, Louis Sullivan, promoted an overriding precept to architectural design: "Form follows function".

While the notion that structural and aesthetic considerations should be entirely subject to functionality was met with both popularity and skepticism, it had the effect of introducing the concept of "function" in place of Vitruvius' "utility". "Function" came to be seen as encompassing all criteria of the use, perception and enjoyment of a building, not only practical but also aesthetic, psychological and cultural.

Sydney Opera House, Australiadesigned by Jørn Utzon.

Nunzia Rondanini stated, "Through its aesthetic dimension architecture goes beyond the functional aspects that it has in common with other human sciences. Through its own particular way of expressing values, architecture can stimulate and influence social life without presuming that, in and of itself, it will promote social development.' To restrict the meaning of (architectural) formalism to art for art's sake is not only reactionary; it can also be a purposeless quest for perfection or originality which degrades form into a mere instrumentality".

Among the philosophies that have influenced modern architects and their approach to building design are rationalism, empiricism,structuralism, poststructuralism, and phenomenology.In the late 20th century a new concept was added to those included in the compass of both structure and function, the consideration ofsustainability, hence sustainable architecture. To satisfy the contemporary ethos a building should be constructed in a manner which is environmentally friendly in terms of the production of its materials, its impact upon the natural and built environment of its surrounding area and the demands that it makes upon non-sustainable power sources for heating, cooling, water and waste management and lighting

HISTORY

ORIGINS AND VERNACULAR ARCHITECTURE

Fig. Vernacular architecture in Norway.

Building first evolved out of the dynamics between needs (shelter, security, worship, etc.) and means (available building materials and attendant skills). As human cultures developed and knowledge began to be formalized through oral traditions and practices, building became a craft, and "architecture" is the name given to the most highly formalized and respected versions of that craft.

It is widely assumed that architectural success was the product of a process of trial and error, with progressively less trial and more replication as the results of the process proved increasingly satisfactory. What is termed vernacular architecture continues to be produced in many parts of the world. Indeed, vernacular buildings make up most of the built world that people experience every day. Early human settlements were mostly rural. Due to a surplus in production the economy began to expand resulting in urbanization thus creating urban areas which grew and evolved very rapidly in some cases, such as that of Çatal Höyük in Anatolia and Mohenjo Daro of the Indus Valley Civilization in modern-day Pakistan.

Fig. The Pyramids at Giza in Egypt.

ANCIENT ARCHITECTURE

In many ancient civilizations, such as those of Egypt and Mesopotamia, architecture and urbanism reflected the constant engagement with the divine and the supernatural, and many ancient cultures resorted to monumentality in architecture to represent symbolically the political power of the ruler, the ruling elite, or the state itself.

The architecture and urbanism of the Classical civilizations such as the Greek and the Roman evolved from civic ideals rather than religious or empirical ones and new building types emerged. Architectural "style" developed in the form of the Classical orders.

Texts on architecture have been written since ancient time. These texts provided both general advice and specific formal prescriptions or canons. Some examples of canons are found in the writings of the 1st-century BCE Roman military engineer Vitruvius. Some of the most important early examples of canonic architecture are religious.

Fig. Kinkaku-ji (Golden Pavilion), Kyoto, Japan

ASIAN ARCHITECTURE

Early Asian writings on architecture include the *Kao Gong Ji* of China from the 7th–5th centuries BCE; the Shilpa Shastras of ancientIndia and Manjusri Vasthu Vidya Sastra of Sri Lanka. The architecture of different parts of Asia developed along different lines from that of Europe; Buddhist, Hindu and Sikh architecture each having different characteristics. Buddhist architecture, in particular, showed great regional diversity. Hindu temple architecture, which developed around the 3rd century BCE, is governed by concepts laid down in the Shastras, and is concerned with expressing the macrocosm and the microcosm. In many Asian countries, pantheistic religion led to architectural forms that were designed specifically to enhance the natural landscape.

Fig. The Taj Mahal (1632–1653), in India

INDIAN ARCHITECTURE

Islamic architecture began in the 7th century CE, incorporating architectural forms from the ancient Middle East and Byzantium, but also developing features to suit the religious and social needs of the society. Examples can be found throughout the Middle East, North Africa, Spain and the Indian Sub-continent. The widespread application of the pointed arch was to influence European architecture of the Medieval period.

THE MEDIEVAL BUILDER

Fig. Notre Dame de Paris, France

In Europe during the Medieval period, guilds were formed by craftsmen to organize their trades and written contracts have survived, particularly in relation to ecclesiastical buildings. The role of architect was usually one with that of master mason, or *Magister lathomorum* as they are sometimes described in contemporary documents.The major architectural undertakings were the buildings of abbeys and cathedrals. From about 900 CE onwards, the movements of both clerics and tradesmen carried architectural knowledge across Europe, resulting in the pan-European styles Romanesque and Gothic.

Fig. La Rotonda (1567), Italy by Palladio

RENAISSANCE AND THE ARCHITECT

In Renaissance Europe, from about 1400 onwards, there was a revival of Classical learning accompanied by the development of Renaissance Humanism which placed greater emphasis on the role of the individual in society than had been the case during the Medieval period. Buildings were ascribed to specific architects –Brunelleschi, Alberti, Michelangelo, Palladio – and the cult of the individual had begun. There was still no dividing line between artist,architect and engineer, or any of the related vocations, and the appellation was often one of regional preference.A revival of the Classical style in architecture was accompanied by a burgeoning of science and engineering which affected the proportions and structure of buildings. At this stage, it was still possible for an artist to design a bridge as the level of structural calculations involved was within the scope of the generalist.

EARLY MODERN AND THE INDUSTRIAL AGE

Fig. Paris Opera by Charles Garnier(1875), France

With the emerging knowledge in scientific fields and the rise of new materials and technology, architecture and engineering began to separate, and the architect began to concentrate on aesthetics and the humanist aspects, often at the expense of technical aspects of building design. There was also the rise of the "gentleman architect" who usually dealt with wealthy clients and concentrated predominantly on visual qualities derived usually from historical prototypes, typified by the many country houses of Great Britain that were created in the Neo- Gothic or Scottish Baronial styles. Formal architectural training in the 19th century, for example at Ecole des Beaux Arts in France, gave much emphasis to the production of beautiful drawings and little to context and feasibility. Effective architects generally received their training in the offices of other architects, graduating to the role from draughtsmen or clerks.

Meanwhile, the Industrial Revolution laid open the door for mass production and consumption. Aesthetics became a criterion for the middle class as ornamented products, once within the province of expensive craftsmanship, became cheaper under machine production.

Vernacular architecture became increasingly ornamental. House builders could use current architectural design in their work by combining features found in pattern books and architectural journals.

MODERNISM AND REACTION

Fig. The Bauhaus Dessau architecture department from 1925 by Walter Gropius

Around the beginning of the 20th century, a general dissatisfaction with the emphasis on revivalist architecture and elaborate decoration gave rise to many new lines of thought that served as precursors to Modern Architecture. Notable among these is the Deutscher Werkbund, formed in 1907 to produce better quality machine made objects. The rise of the profession of industrial design is usually placed here. Following this lead, the Bauhaus school, founded in Weimar, Germany in 1919, redefined the architectural bounds prior set throughout history, viewing the creation of a building as the ultimate synthesis—the apex—of art, craft, and technology. When Modern architecture was first practiced, it was an avant-garde movement with moral, philosophical, and aesthetic underpinnings.

Immediately after World War I, pioneering modernist architects sought to develop a completely new style appropriate for a new post-war social and economic order, focused on meeting the needs of the middle and working classes.

They rejected the architectural practice of the academic refinement of historical styles which served the rapidly declining aristocratic order. The approach of the Modernist architects was to reduce buildings to pure forms, removing historical references and ornament in Favour of functionalist details. Buildings displayed their functional and structural elements, exposing steel beams and concrete surfaces instead of hiding them behind decorative forms.

Fig. Fallingwater, Organic architectureby Frank Lloyd Wright.

Architects such as Frank Lloyd Wright developed Organic architecture, in which the form was defined by its environment and purpose, with an aim to promote harmony between human habitation and the natural world with prime examples being Robie House andFallingwater.

Fig. The Crystal Cathedral, California, byPhilip Johnson (1980)

Architects such as Mies van der Rohe, Philip Johnson and Marcel Breuer worked to create beauty based on the inherent qualities of building materials and modern construction techniques, trading traditional historic forms for simplified geometric forms, celebrating the new means and methods made possible by the Industrial Revolution, including steel-frame construction, which gave birth to high-rise superstructures.

By mid-century, Modernism had morphed into the International Style, an aesthetic epitomized in many ways by the Twin Towers of New York's World Trade Center.

Many architects resisted Modernism, finding it devoid of the decorative richness of ornamented styles and as the founders of that movement lost influence in the late 1970s, Postmodernismdeveloped as a reaction against its austerity. Postmodernism viewed Modernism as being too extreme and even harsh in regards to design.

Instead, Postmodernists combined Modernism with older styles from before the 1900s to form a middle ground. Robert Venturi's contention that a "decorated shed" (an ordinary building which is functionally designed inside and embellished on the outside) was better than a "duck" (an ungainly building in which the whole form and its function are tied together) gives an idea of these approaches.

ARCHITECTURE TODAY

Fig. Postmodern design at Gare do Oriente, Lisbon, Portugal, by Santiago Calatrava.

Since the 1980s, as the complexity of buildings began to increase (in terms of structural systems, services, energy and technologies), the field of architecture became multi-disciplinary with specializations for each project type, technological expertise or project delivery methods. In addition, there has been an increased separation of the 'design' architect from the 'project' architect who ensures that the project meets the required standards and deals with matters of liability.

The preparatory processes for the design of any large building have become increasingly complicated, and require preliminary studies of such matters as durability, sustainability, quality, money, and compliance with local laws. A large structure can no longer be the design of one person but must be the work of many.

Modernism and Postmodernism, have been criticised by some members of the architectural profession who feel that successful architecture is not a personal philosophical or aesthetic pursuit by individualists; rather it has to consider everyday needs of people and use technology to create liveable environments, with the design process being informed by studies of behavioral, environmental, and social sciences.

Fig. Green roof planted with native species at L'Historial de la Vendée, a new museum in western France.

Environmental sustainability has become a mainstream issue, with profound affect on the architectural profession. Many developers, those who support the financing of buildings, have become educated to encourage the facilitation of environmentally sustainable design, rather than solutions based primarily on immediate cost.

Major examples of this can be found in Passive solar building design, greener roof designs, biodegradable materials, and more attention to a structure's energy usage. This major shift in architecture has also changed architecture schools to focus more on the environment. Sustainability in architecture was pioneered by Frank Lloyd Wright, in the 1960s by Buckminster Fuller and in the 1970s by architects such as Ian McHarg and Sim Van der Ryn in the US and Brenda and Robert Vale in the UK and New Zealand.

There has been an acceleration in the number of buildings which seek to meet green building sustainable design principles. Sustainable practices that were at the core of vernacular architecture increasingly provide inspiration for environmentally and socially sustainable contemporary techniques. The U.S. Green Building Council's LEED (Leadership in Energy and Environmental Design) rating system has been instrumental in this.

Concurrently, the recent movements of New Urbanism and New Classical Architecture promote a sustainable approach towards construction, that appreciates and develops smart growth, architectural tradition and classical design. This in contrast to modernist and globally uniform architecture, as well as leaning against solitary housing estates and suburban sprawl.

HISTORY BACKGROUND OF ARCHITECTURE ENGINEERING

Fig. Dome seen from the Giotto's Campanile

Fig. A view of Chuo-ku, Osaka, Japan showing buildings of a modern Asian city, ranging from the medieval Osaka Castle to skyscrapers

The history of architecture traces the changes in architecture through various traditions, regions, overarching stylistic trends, and dates. The branches of architecture are civil, sacred, naval, military, and landscape architecture.

NEO-LITHIC ARCH

Fig. Excavated dwellings at Skara Brae

Neo-lithics architecture is the architecture of the Neo-lithic period. In Southwest Asia, Neo-lithic cultures appear soon after 10,000 BC, initially in the Levant (Pre-Pottery Neo-lithic A and Pre-Pottery Neo-lithic B) and from there spread eastwards and westwards.

There are early Neo-lithic cultures in Southeast Anatolia, Syria and Iraq by 8000 BC, and food-producing societies first appear in southeast Europe by 7000 BC, and Central Europe by c. 5500 BC (of which the earliest cultural complexes include the Starèevo-Koros (Cris),Linearbandkeramic, and Vinèa). With very small exceptions (a few copper hatchets and spear heads in the Great Lakes region), the people of the Americas and the Pacific remained at the Neo-lithic level of technology up until the time of European contact.

The Neo-lithic people in the Levant, Anatolia, Syria, northern Mesopotamia and Central Asia were great builders, utilizing mud-brick to construct houses and villages. At Çatalhöyük, houses were plastered and painted with elaborate scenes of humans and animals. The Mediterranean Neo-lithic cultures of Malta worshiped in megalithic temples.

In Europe, long houses built from wattle and daub were constructed. Elaborate tombs for the dead were also built.

These tombs are particularly numerous in Ireland, where there are many thousands still in existence. Neo-lithic people in the British Isles built long barrows and chamber tombs for their dead and causewayed camps, henges flint mines and cursus monuments.

ISLAMIC ARCHITECTURE

Fig. Moorish architecture: TheGreat Mosque of Kairouan inTunisia

Islamic architecture has encompassed a wide range of both secular and religious architecture styles from the foundation of Islam to the present day, influencing the design and construction of buildings and structures within the sphere of Islamic culture.

Fig. Ottoman architecture:Sultan Ahmed Mosque, Istanbul, Turkey.

The wide spread and long history of Islam has given rise to many local architectural styles, including Abbasid, Persian, Moorish, Timurid, Ottoman, Fatimid, Mamluk, Mughal,Indo-Islamic, Sino-Islamic and Afro-Islamic architecture.

Fig. Persian architecture:Sheikh Lotf Allah Mosque, Isfahan, Iran.

Some distinctive structures in Islamic architecture are mosques, tombs, palaces and forts, although Islamic architects have of course also applied their distinctive design precepts to domestic architecture.Notable Islamic architectural types include the early Abbasid buildings, T-type mosques, and the central-dome mosques of Anatolia.- Also, Islamic architecture also discourages illustrations of anything living, such as animals and humans.

Various regional styles of medieval Islamic architecture, as show in religious structures (from west to east)

Fig. Mughal architecture:Badshahi Mosque, Pakistan

AFRICAN ARCHITECTURE

Fig. The conical tower inside the Great Enclosure in Great Zimbabwe, a medieval city built by a prosperous culture.

Great Zimbabwe is the largest medieval city in sub-Saharan Africa. By the late nineteenth century, most buildings reflected the fashionable European eclecticism and pastisched Mediterranean, or even Northern European, styles.

In the Western Sahel region, Islamic influence was a major contributing factor to architectural development from the later ages of theKingdom of Ghana. At Kumbi Saleh, locals lived in domed-shaped dwellings in the king's section of the city, surrounded by a great enclosure.

Traders lived in stone houses in a section which possessed 12 beautiful mosques, as described by al-bakri, with one centered on Friday prayer. The king is said to have owned several mansions, one of which was sixty-six feet long, forty-two feet wide, contained seven rooms, was two stories high, and had a staircase; with the walls and chambers filled with sculpture and painting.Sahelian architecture initially grew from the two cities of Djenné and Timbuktu. The Sankore Mosque in Timbuktu, constructed from mud on timber, was similar in style to the Great Mosque of Djenné.

The rise of kingdoms in the West African coastal region produced architecture which drew on indigenous traditions, utilizing wood. The famed Benin City, destroyed by the Punitive Expedition, was a large complex of homes in coursed clay, with hipped roofs of shingles or palm leaves. The Palace had a sequence of ceremonial rooms, and was decorated with brass plaques.

SOUTHERN ASIAN ARCHITECTURE

INDIAN ARCHITECTURE

Fig. Chennakesava Temple, Belur in Karnataka, India.

Indian architecture encompasses a wide variety of geographically and historically spread structures, and was transformed by the history of the Indian subcontinent. The result is an evolving range of architectural production that, although it is difficult to identify a single representative style, Non-etheless retains a certain amount of continuity across history.

The diversity of Indian culture is represented in its architecture. It is a blend of ancient and varied native traditions, with building types, forms and technologies from West and Central Asia, as well as Europe.

Architectural styles range from Hindu temple architecture to Islamic architecture to western classical architecture tomodern and post-modern architecture. India's Urban Civilization is traceable to Mohenjodaro and Harappa, now in Pakistan.

From then on, Indian architecture and civil engineering continued to develop, and was manifestated temples, palaces and forts across the Indian subcontinent and neighbouring regions. Architecture and civil engineering was known as *sthapatya-kala*, literally "the art of constructing".

The temples of Aihole and Pattadakal are the earliest known examples of Hindu temples. There are numerous Hindu as well as Buddhist temples that are known as excellent examples of Indian rock-cut architecture. According to J.J. O'Connor and E. F. Robertson, the*Sulbasutras* were appendices to the Vedas giving *rules for constructing altars*. "They contained quite an amount of geometrical knowledge, but the mathematics was being developed, not for its own sake, but purely for practical religious purposes."

Fig. The Hall of Private Audience at Fatehpur Sikri in Uttar Pradesh, India, an early example of the architecture of the Mughal Empire.

During the Kushan Empire and Mauryan Empire, Indian architecture and civil engineering reached regions like Baluchistan andAfghanistan. Statues of Buddha were cut out, covering entire mountain cliffs, like in Buddhas of Bamyan, Afghanistan. Over a period of time, ancient Indian art of construction blended with Greek styles and spread to Central Asia. It includes the architecture of various dynasties, such as Hoysala architecture, Vijayanagara architecture and Western Chalukya architecture.The Church of St. Anne which is cast in the Indian Baroque Architectural style under the orientation of the most eminent architects of the time. It is a prime example of the blending of traditional Indian styles with western European architectural styles.

BUDDHIST ARCHITECTURE

Fig. Stupa at the top of Borobudur, Java, Indonesia

EASTERN ASIA

CHINESE ARCHITECTURE

From the Neo-lithic era Longshan Culture and Bronze Age era Erlitou culture, the earliest rammed earth fortifications exist, with evidence oftimber architecture. The subterranean ruins of the palace at Yinxu dates back to the

Shang Dynasty (c. 1600 BC–1046 BC). In historic China, architectural emphasis was laid upon the horizontal axis, in particular the construction of a heavy platform and a large roof that floats over this base, with the vertical walls not as well emphasized.

Fig. The Iron Pagoda of Kaifeng, built in 1049 during the Song Dynasty.

This contrasts Western architecture, which tends to grow in height and depth. Chinese architecture stresses the visual impact of the width of the buildings.

The deviation from this standard is the tower architecture of the Chinese tradition, which began as a native tradition and was eventually influenced by the Buddhist building for housing religious sutras — the stupa — which came from India.

Ancient Chinese tomb model representations of multiple story residential towers and watchtowers date to the Han Dynasty (202 BC–220 AD). However, the earliest extant Buddhist Chinese pagoda is the Songyue Pagoda, a 40 m (131 ft) tall circular-based brick tower built in Henan province in the year 523 AD. From the 6th century onwards, stone-based structures become more common, while the earliest are from stone and brick arches found in Han Dynasty tombs. The Zhaozhou Bridge built from 595 to 605 AD is China's oldest extant stone bridge, as well as the world's oldest fully stone open-spandrel segmental arch bridge.

Fig. Inside the Forbidden City- an example of Chinese architecture from the 15th century.

The vocational trade of architect, craftsman, and engineer was not as highly respected in premodern Chinese society as the scholar-bureaucrats who were drafted into the government by the civil service examination system. Much of the knowledge about early Chinese architecture was passed on from one tradesman to his son or associative apprentice. However, there were several early treatises on architecture in China, with encyclopedic information on architecture dating back to the Han Dynasty.

The height of the classical Chinese architectural tradition in writing and illustration can be found in the *Yingzao Fashi*, a building manual written by 1100 and published by Lie Jie (1065–1110) in 1103. In it there are numerous and meticulous illustrations and diagrams showing the assembly of halls and building components, as well as classifying structure types and building components.

There were certain architectural features that were reserved solely for buildings built for the Emperor of China. One example is the use of yellow roof tiles; yellow having been the Imperial Colour, yellow roof tiles still adorn most of the buildings within the Forbidden City. TheTemple of Heaven, however, uses blue roof tiles to symbolize the sky.

The roofs are almost invariably supported by brackets, a feature shared only with the largest of religious buildings. The wooden columns of the buildings, as well as the surface of the walls, tend to be red in colour.

Many current Chinese architectural designs follow post-modern and western styles.

KOREAN ARCHITECTURE

Korean architecture has a long history of 8,000 years. It has its own cultural identity different from Chinese or Japanese architecture. The basic construction form is more or less similar to Eastern Asian building system. From a technical point of view, buildings are structured vertically and horizontally. A construction usually rises from a stone subfoundation to a curved roof covered with tiles, held by a console structure and supported on posts; walls are made of earth (adobe) or are sometimes totally composed of movable wooden doors. Architecture is built according to the k'an unit, the distance between two posts (about 3.7 meters), and is designed so that there is always a transitional space between the "inside" and the "outside."

Fig. Throne Hall of GyeongbokgungPalace, Seoul, South Korea

The console, or bracket structure, is a specific architectonic element that has been designed in various ways through time. If the simple bracket system was already in use under the Goguryeo kingdom (37 BCE–668 CE)—in palaces in Pyongyang, for instance—a curved version, with brackets placed only on the column heads of the building, was elaborated during the early Koryo dynasty (918–1392).

The Amita Hall of the Pusok temple in Antong is a good example. Later on (from the mid-Koryo period to the early Choson dynasty), a multiple-bracket system, or an inter-columnar-bracket set system, was developed under the influence of Mongol's Yuan dynasty (1279–1368). In this system, the consoles were also placed on the transverse horizontal beams. Seoul's Namtaemun Gate Namdaemun, Korea's foremost national treasure, is perhaps the most symbolic example of this type of structure

In the mid-Choson period, the winglike bracket form appeared (one example is the Yongnyongjon Hall of Jongmyo, Seoul), which is interpreted by many scholars as an example of heavy Confucian influence in Joseon Korea, which emphasized simplicity and modesty in such shrine buildings. Only in buildings of importance like palaces or sometimes temples (Tongdosa, for instance) were the multicluster brackets still used. Confucianism also led to more sober and simple solutions.

JAPANESE ARCHITECTURE

Japanese architecture has as long a history as any other aspect of Japanese culture. Influenced heavily by Chinese and Korean architecture, it also shows a number of important differences and aspects which are uniquely Japanese.

Two new forms of architecture were developed in medieval Japan in response to the militaristic climate of the times: the castle, a defensive structure built to house a feudal lord and his soldiers in times of trouble; and the *shoin*, a reception hall and private study area designed to reflect the relationships of lord and vassal within a feudal society.

Fig. View of Himeji Castle from Nishi-no-maru

Because of the need to rebuild Japan after World War II, major Japanese cities contain numerous examples of modern architecture. Japan played some role in modern skyscraper design, because of its long familiarity with the cantilever principle to support the weight of heavy tiled temple roofs. New city planning ideas based on the principle of layering or cocooning around an inner space (oku), a Japanese spatial concept that was adapted to urban needs, were adapted during reconstruction. Modernism became increasingly popular in architecture in Japan starting in the 1950s.

PRE-COLUMBIAN

Fig. Overview of the central plaza of the Mayan city of Palenque(Chiapas, Mexico), a fine example of Classic period Mesoamerican architecture.

MESOAMERICAN ARCHITECTURE

Mesoamerican architecture is the set of architectural traditions produced by pre-Columbian cultures and civilizations of Mesoamerica, (such as the Olmec,

Maya, and Aztec) traditions which are best known in the form of public, ceremonial and urban monumental buildings and structures.

The distinctive features of Mesoamerican architecture encompass a number of different regional and historical styles, which however are significantly interrelated.

These styles developed throughout the different phases of Mesoamerican history as a result of the intensive cultural exchange between the different cultures of the Mesoamerican culture area through thousands of years. Mesoamerican architecture is mostly noted for its pyramids which are the largest such structures outside of Ancient Egypt.

INCAN ARCHITECTURE

Fig. View of Machu Picchu

Incan architecture consists of the major construction achievements developed by the Incas. The Incas developed an extensive road systemspanning most of the western length of the continent. Inca rope bridges could be considered the world's first suspension bridges. Because the Incas used no wheels (It would have been impractical for the terrain) or horses, they built their roads and bridges for foot and pack-llama traffic.

Much of present day architecture at the former Inca capital Cuzco shows both Incan and Spanish influences. The famous lost city Machu Picchu is the best surviving example of Incan architecture. Another significant site is Ollantaytambo. The Inca were sophisticated stone cutters whose masonry used no mortar.

ANCIENT ARCHITECTURE OF NORTH AMERICA

Inside what is the present-day United States, the Mississippians and the Pueblo created substantial public architecture. TheMississippian culture was among the mound-building peoples, noted for construction of large earthen platform mounds.

Impermanent buildings, which were often architecturally unique from region to region, continue to influence American architecture today. In his summary,

"The World of Textiles", North Carolina State's Tushar Ghosh provides one example: the Denver International Airport's roof is a fabric structure that was influenced by and/or resembles the tipis of local cultures.

Fig. Cliff Palace of Mesa Verde, in Colorado, United States, created by theAncient Pueblo Peoples.

In writing about Evergreen State College, Lloyd Vaughn lists an example of very different native architecture that also influenced contemporary building: the Native American Studies Programme is housed in a modern-day longhouse derived from pre-Columbian Pacific Northwest architecture.

EUROPE TO 1600

MEDIEVALARCHITECTURE

Surviving examples of medieval secular architecture mainly served for Defence. Castles and fortified walls provide the most notable remaining non-religious examples of medieval architecture. Windows gained a cross-shape for more than decorative purposes: they provided a perfect fit for a crossbowman to safely shoot at invaders from inside.Crenellation walls (battlements) provided shelters for archers on the roofs to hide behind when not shooting.

PRE-ROMANESQUE

Western European architecture in the Early Middle Ages may be divided into Early Christian and Pre-Romanesque, including Merovingian, Carolingian, Ottonian, and Asturian. While these terms are problematic, they Non-etheless serve adequately as entries into the era. Considerations that enter into histories of each period include Trachtenberg's "historicising" and "modernising" elements, Italian versus northern, Spanish, and Byzantine elements, and especially the religious and political maneuverings between kings, popes, and various ecclesiastic officials.

ROMANESQUE

Fig. Karlštejn is a large Gothic castle founded 1348 by Charles IV, Holy Roman Emperor.

Romanesque, prevalent in medieval Europe during the 11th and 12th centuries, was the first pan-European style since Roman Imperial architecture and examples are found in every part of the continent. The term was not contemporary with the art it describes, but rather, is an invention of modern scholarship based on its similarity to Roman architecture in forms and materials. Romanesque is characterized by a use of round or slightly pointed arches, barrel vaults, and cruciform piers supporting vaults.

GOTHIC

Fig. Norse architecture: Borgund Stave Church, Norway

The various elements of Gothic architecture emerged in a number of 11th- and 12th-century building projects, particularly in the Île de France area, but were first combined to form what we would now recognise as a distinctively

Gothic style at the 12th century abbey church of Saint-Denis in Saint-Denis, near Paris.

Fig. Romanesque architecture:
Cathedral of Santa Maria Maior, Lisbon, Portugal

Verticality is emphasized in Gothic architecture, which features almost skeletal stone structures with great expanses of glass, pared-down wall surfaces supported by external flying buttresses, pointed arches using the ogiveshape, ribbed stone vaults, clustered columns, pinnacles and sharply pointed spires. Windows contain beautiful stained glass, showing stories from the Bible and from lives of saints.

Such advances in design allowed cathedrals to rise taller than ever, and it became something of an inter-regional contest to build a church as high as possible.

Fig. Gothic architecture: Notre-Dame de Chartres, France (1194–1260) -

RENAISSANCE ARCHITECTURE

Fig. St. Peter's Basilica in Rome.

The Renaissance often refers to the Italian Renaissance that began in the 14th century, but recent research has revealed the existence of similar movements around Europe before the 15th century; consequently, the term "Early Modern" has gained popularity in describing this cultural movement. This period of cultural rebirth is often credited with the restoration of scholarship in the Classical Antiquities and the absorption of new scientific and philosophical knowledge that fed the arts.

The development from Medieval architecture concerned the way geometry mediated between the intangibility of light and the tangibility of the material as a way of relating divine creation to mortal existence.

This relationship was changed in some measure by the invention of Perspective which brought a sense of infinity into the realm of human comprehension through the new representations of the horizon, evidenced in the expanses of space opened up in Renaissance painting, and helped shape new humanist thought.

Perspective represented a new understanding of space as a universal, *a priori* fact, understood and controllable through human reason.

Renaissance buildings therefore show a different sense of conceptual clarity, where spaces were designed to be understood in their entirety from a specific fixed viewpoint.

The power of Perspective to universally represent reality was not limited to *describing* experiences, but also allowed it to anticipate experience itself by projecting the image back into reality.

The Renaissance spread to France in the late 15th century, when Charles VIII returned in 1496 with several Italian artists from his conquest of Naples. Renaissance chateaux were built in the Loire Valley, the earliest example being the Château d'Amboise, and the style became dominant under Francis I (1515–47).

The Château de Chambord is a combination of Gothic structure and Italianate ornament, a style which progressed under architects such as Sebastiano Serlio, who was engaged after 1540 in work at the Château de Fontainebleau.

Architects such as Philibert Delorme, Androuet du Cerceau, Giacomo Vignola, and Pierre Lescot, were inspired by the new ideas.

The southwest interior facade of the Cour Carree of the Louvre in Paris was designed by Lescot and covered with exterior carvings by Jean Goujon. Architecture continued to thrive in the reigns of Henri II and Henri III.

In England the first great exponent of Renaissance architecture was Inigo Jones (1573–1652), who had studied architecture in Italy where the influence of Palladio was very strong.

Jones returned to England full of enthusiasm for the new movement and immediately began to design such buildings as the Queen's House at Greenwich in 1616 and the Banqueting

House at Whitehall three years later.

These works with their clean lines and symmetry, were revolutionary in a country still enamoured with mullion windows, crenellations and turrets.

EUROPEAN AND COLONIAL ARCHITECTURE

Fig. Sicilian Baroque: Basilica della Collegiata, Catania, Sicily, Italy.

With the rise of various European colonial empires from the 16th century onward through the early 20th century, the new stylistic trends of Europe were exported to or adopted by locations around the world, often evolving into new regional variations.

BAROQUE ARCHITECTURE

The periods of Mannerism and the Baroque that followed the Renaissance signaled an increasing anxiety over meaning and representation. Important developments in science and philosophy had separated mathematical representations of reality from the rest of culture, fundamentally changing the way humans related to their world through architecture. It would reach its most extreme and embellished development under the decorative tastes of Rococo.

RETURN TO CLASSICISM

In the late 17th and 18th centuries, the works and theories of Andrea Palladio (from 16th-century Venice) would again be interpreted and adopted in England, spread by the English translation of his I Quattro Libri dell'Architettura, and pattern books such as *Vitruvius Brittanicus* by Colen Campbell. This Palladian architecture and continued classical imagery would in turn go on to influence Thomas Jefferson and other early architects of the United States in their search for a new national architecture.

By the mid-18th century, there tended to be more restrained decoration and usage of authentic classical forms than in the Baroque, informed by increased visitation to classical ruins as part of the Grand Tour, coupled with the excavations of Pompeii and Herculaneum.

Federal-style architecture is the name for the classicizing architecture built in North America between c. 1780 and 1830, and particularly from 1785 to 1815. This style shares its name with its era, the Federal Period. The term is also used in association with furniture design in the United States of the same time period. The style broadly corresponds to the middle-class classicism of Biedermeier style in the German-speaking lands, Regency style in Britain and to the French Empire style.

REVIVALISM AND ORIENTALISM

The 19th Century was dominated by a wide variety of stylistic revivals, variations, and interpretations.

BEAUX-ARTS ARCHITECTURE

Beaux-Arts architecture denotes the academic classical architectural style that was taught at the École des Beaux Arts in Paris. The*style "Beaux-Arts"* is above all the cumulative product of two and a half centuries of instruction under the authority, first of the Académie royale d'architecture, then, following the Revolution, of the Architecture section of the Académie des Beaux-Arts. The organization under the Ancien Régime of the competition for the Grand Prix de Rome in architecture, offering a chance to study in Rome, imprinted its codes and esthetic on the course of instruction, which culminated during the Second Empire (1850–1870) and the Third Republic that followed. The style of instruction that produced Beaux-Arts architecture continued without a major renovation until 1968.

Fig. Palais Garnier is a cornerpiece of Beaux-Arts architecture characterized by Émile Zola as "the opulent bastard of all styles".

ART NOUVEAU

Around 1900 a number of architects around the world began developing new architectural solutions to integrate traditional precedents with new social demands and technological possibilities. The work of Victor Horta and Henry van de Velde in Brussels, Antoni Gaudí in Barcelona, Otto Wagner in Vienna and Charles Rennie Mackintosh in Glasgow, among many others, can be seen as a common struggle between old and new.

EARLY MODERN ARCHITECTURE

Early Modern architecture began with a number of building styles with similar characteristics, primarily the simplification of form and the elimination of ornament, that first arose around 1900. By the 1940s these styles had largely consolidated and been identified as the International Style.

The exact characteristics and origins of modern architecture are still open to interpretation and debate. An important trigger appears to have been the maxim credited to Louis Sullivan: "form follows function". Functionalism, in architecture, is the principle that architects should design a building based on the purpose of that building.

This statement is less self-evident than it first appears, and is a matter of confusion and controversy within the profession, particularly in regard to modern architecture.

EXPRESSIONIST ARCHITECTURE

Fig. Goetheanum by Rudolf Steiner in 1923

Expressionist architecture was an architectural movement that developed in Northern Europe during the first decades of the 20th century in parallel with the expressionist visual and performing arts.

The style was characterised by an early-modernist adoption of novel materials, formal innovation, and very unusual massing, sometimes inspired by natural biomorphic forms, sometimes by the new technical possibilities offered by the mass production of brick, steel and especially glass. Many expressionist architects fought in World War I and their experiences, combined with the political turmoil and social upheaval that followed the German Revolution of 1919, resulted in a utopian outlook and a romantic socialist agenda. Economic conditions severely limited the number of built commissions between 1914 and the mid-1920s, resulting in many of the most important expressionist works remaining as projects on paper, such as Bruno Taut's *Alpine Architecture* and Hermann Finsterlin's *Formspiels*.

Ephemeral exhibition buildings were numerous and highly significant during this period. Scenography for theatre and films provided another outlet for the expressionist imagination, and provided supplemental incomes for designers attempting to challenge conventions in a harsh economic climate.

INTERNATIONAL STYLE

The International style was a major architectural trend of the 1920s and 1930s. The term usually refers to the buildings and architects of the formative decades of modernism, before World War II. The term had its origin from the name of a book by Henry-Russell Hitchcockand Philip Johnson which identified, categorised and expanded upon characteristics common to modernism across the world. As a result, the focus was more on the stylistic aspects of modernism. The basic design principles of the International Style thus constitute part ofmodernism.

Fig. The Glass Palace, a celebration of transparency, in Heerlen, The Netherlands (1935)

The ideas of Modernism were developed especially in what was taught at the German Bauhaus School in Weimar (from 1919), Dessau(between 1926–32) and finally Berlin between 1932–33, under the leadership first of its founder Walter Gropius, then Hannes Meyer, and finally Ludwig Mies van der Rohe. Modernist theory in architecture resided in the attempt to bypass the question of what style a building should be built in, a concern that had overshadowed 19th-century architecture, and the wish to reduce form to its most minimal expression of structure and function.

In the USA, Philip Johnson and Henry-Russell Hitchcock treated this new phenomenon in 1931 as if it represented a new style - the International Style, thereby misrepresenting its primary mission as merely a matter of eliminating traditional ornament. The core effort to pursue Modern architecture as an abstract, scientific programme was more faithfully carried forward in Europe, but issues of style always overshadowed its stricter and more puritan goals, not least in the work of Le Corbusier.

CONTEMPORARY ARCHITECTURE

MODERN ARCHITECTURE

Modern architecture is generally characterized by simplification of form and creation of ornament from the structure and theme of the building. It is a term applied to an overarching movement, with its exact definition and scope varying widely.

Modern architecture has continued into the 21st century as a contemporary style, especially for corporate office buildings. In a broader sense, modern architecture began at the turn of the 20th century with efforts to reconcile the principles underlying architectural design with rapid technological advancement and the modernization of society. It would take the form of numerous movements, schools of design, and architectural styles, some in tension with one another, and often equally defying such classification.

CRITICAL REGIONALISM

Fig. The Sydney Opera House - designed to evoke the sails of yachts in Sydney harbour

Critical regionalism is an approach to architecture that strives to counter the placelessness and lack of meaning in Modern architectureby using contextual forces to give a sense of place and meaning. The term critical regionalism was first used by Alexander Tzonis andLiane Lefaivre and later more famously by Kenneth Frampton.

Frampton put forth his views in *"Towards a Critical Regionalism: Six points of an architecture of resistance."* He evokes Paul Ricœur's question of "how to become modern and to return to sources; how to revive an old, dormant civilization and take part in universal civilization". According to Frampton, critical regionalism should adopt modern architecture critically for its universal progressive qualities but at the same time should value responses particular to the context. Emphasis should be on topography, climate, light, tectonic form rather than scenography and the tactile sense rather than the visual. Frampton draws from phenomenology to supplement his arguments.

POSTMODERN ARCHITECTURE

Postmodern architecture is an international style whose first examples are generally cited as being from the 1950s, and which continues to influence present-day architecture. Postmodernity in architecture is generally thought to be heralded by the return of "wit, ornament and reference" to architecture in response to the formalism of the International Style of modernism. As with many cultural movements, some of postmodernism's most pronounced and visible ideas can be seen in architecture. The functional and formalized shapes and spaces of the modernist movement are replaced by unapologetically diverse aesthetics: styles collide, form is adopted for its own sake, and new ways of viewing familiar styles and space abound.

Fig. 1000 de La Gauchetière, with ornamented and strongly defined top, middle and bottom. Contrast with the modernist Seagram Building and Torre Picasso.

Classic examples of modern architecture are the Lever House and the Seagram Building in commercial space, and the architecture of Frank Lloyd Wright or the Bauhaus movement in private or communal spaces.

Transitional examples of postmodern architecture are the Portland Building inPortland and the Sony Building (New York City) (originally ATandT Building) in New York City, which borrows elements and references from the past and reintroduces Colour and symbolism to architecture. A prime example of inspiration for postmodern architecture lies along the Las Vegas Strip, which was studied by Robert Venturi in his 1972 book *Learning from Las Vegas* celebrating the strip's ordinary and common architecture. Venturi opined that "Less is a bore", inverting Mies Van Der Rohe's dictum that "Less is more".

Following the postmodern movement, a renaissance of pre-modernist urban and architectural ideals established itself, with New Urbanism and New Classical architecture being prominent movements.

DECONSTRUCTIVIST ARCHITECTURE

Deconstructivism in architecture is a development of postmodern architecture that began in the late 1980s. It is characterized by ideas of fragmentation, non-linear processes of design, an interest in manipulating ideas of a structure's surface or skin, and apparent non-Euclidean geometry, (*i.e.*, non-rectilinear shapes) which serve to distort and dislocate some of the elements of architecture, such as structure and envelope.

Fig. Libeskind's Imperial War Museum North in Manchester comprises three apparently intersecting curved volumes.

The finished visual appearance of buildings that exhibit the many deconstructivist "styles" is characterised by a stimulating unpredictability and a controlled chaos. Important events in the history of the deconstructivist movement include the 1982 Parc de la Villette architectural design competition(especially the entry from Jacques Derrida and Peter Eisenman and Bernard Tschumi's winning entry), the Museum of Modern Art's 1988 *Deconstructivist Architecture* exhibition in New York, organized by Philip Johnson and Mark Wigley, and the 1989 opening of theWexner Center for the Arts in Columbus, designed by Peter Eisenman.

The New York exhibition featured works by Frank Gehry, Daniel Libeskind, Rem Koolhaas, Peter Eisenman, Zaha Hadid, Coop Himmelblau, and Bernard Tschumi. Since the exhibition, many of the architects who were associated with Deconstructivism have distanced themselves from the term. Non-etheless, the term has stuck and has now, in fact, come to embrace a general trend within contemporary architecture.

ARCHITECTURE IN THE 21ST CENTURY

On January 21, 2013 architects began preparations for constructing the world's first 3D-printed building, with completion expected in 2014. An industrial-scale 3D printer will use high strength artificial marble.

Sustainable architecture is an important topic in contemporary architecture, including the trends of New Urbanism, New Classical architecture and Eco-cities.

ARCHITECTURAL THEORY

Architectural theory is the act of thinking, discussing, and writing about architecture. Architectural theory is taught in most architecture schools and is practiced by the world's leading architects. Some forms that architecture theory takes are the lecture or dialogue, the treatise or book, and the paper project or competition entry. Architectural theory is often didactic, and theorists tend to stay close to or work from within schools. It has existed in some form since antiquity, and as publishing became more common, architectural theory gained an increased richness. Books, magazines, and journals published an unprecedented amount of works by architects and critics in the 20th century. As a result, styles and movements formed and dissolved much more quickly than the relatively enduring modes in earlier history. It is to be expected that the use of the internet will further the discourse on architecture in the 21st century.

HISTORY

Antiquity

There is little information or evidence about major architectural theory in antiquity, until the 1st century BCE, with the work of Vitruvius. This does not mean, however, that such works did not exist. Many works never survived antiquity. Vitruvius was a Roman writer, architect, and engineer active in the 1st century BCE. He was the most prominent architectural theorist in the Roman Empire known today, having written *De architectura*, (known today as *The Ten Books of Architecture*), a treatise written of Latin and Greek on architecture, dedicated to the emperor Augustus. Probably written between 27 and 23 BCE, it is the only major contemporary source on classical architecture to have survived.

Divided into ten sections or "books", it covers almost every aspect of Roman architecture, from town planning, materials, decorations, temples, water supplies, etc. It rigorously defines the classical orders of architecture. It also proposes the three fundamental laws that Architecture must obey, in order to be so considered: *firmitas, utilitas, venustas*, translated in the 17th century by Sir Henry Wotton into the English slogan *firmness, commodity and delight* (meaning structural adequacy, functional adequacy, and beauty).

The rediscovery of Vitruvius' work had a profound influence on architects of the Renaissance, adding archaeological underpinnings to the rise of the Renaissance style, which was already under way. Renaissance architects, such as Niccoli, Brunelleschi and Leon Battista Alberti, found in "De Architectura" their rationale for raising their branch of knowledge to a scientific discipline.

Middle Ages

Throughout the Middle Ages, architectural knowledge was passed by transcription, word of mouth and technically in master builders' lodges. Due to the laborious nature of transcription, few examples of architectural theory were penned in this time period.

Most works from this period were theological, and were transcriptions of the bible, so the architectural theories were the notes on structures included therein. The Abbot Suger's *Liber de rebus in administratione sua gestis*, was an architectural document that emerged with gothic architecture. Another was Villard de Honnecourt's portfolio of drawings from about the 1230s.

In Song Dynasty China, Li Jie published the *Yingzao Fashi* in 1103, which was an architectural treatise that codified elements of Chinese architecture.

Renaissance

The first great work of architectural theory of this period belongs to Leon Battista Alberti, *De Re Aedificatoria*, which placed Vitruvius at the core of the most profound theoretical tradition of the modern ages. From Alberti, good architecture is validated through the Vitruvian triad, which defines its purpose. This triplet conserved all its validity until the 19th century. A major transition into the 17th century and ultimately to the phase of Enlightenment was secured through the advanced mathematical and optical research of the celebrated architect and geometer Girard Desargues, with an emphasis on his studies on perspective and projective geometry.

Enlightenment

The Age of the Enlightenment witnessed considerable development in architectural theory on the European continent. New archeological discoveries (such as those of Pompeii and Herculaneum) drove new interest in Classical art and architecture. Thus the term Neo-classicism (exemplified by the writings of Prussian art critic Johann Joachim Winkelmann) arose to designate 18th-century architecture which looked to these new Classical precedents for inspiration in building design.

Major architectural theorists of the Enlightenment include Julien-David Leroy, Abbé Marc-Antoine Laugier, Giovanni Battista Piranesi, Robert Adam, James Stuart, Georg Friedrich Hegel and Nicholas Revett.

19th century

A vibrant strain of Neo-classicism, inherited from Marc-Antoine Laugier's seminal Essai, provided the foundation for two generations of international activity around the core themes of classicism, primitivism and a "return to Nature."

Reaction against the dominance of neo-classical architecture came to the fore in the 1820s with Augustus Pugin providing a moral and theoretical basis for Gothic Revival architecture, and in the 1840s John Ruskin developed this ethos.

The American sculptor Horatio Greenough published the essay *American Architecture* in August 1843 in which he rejected the imitation of old styles of buildings and outlined the functional relationship between architecture and decoration. These theories anticipated the development of Functionalism in modern architecture.

Towards the end of the century, there occurred a blossoming of theoretical activity. In England, Ruskin's ideals underpinned the emergence of the Arts and Crafts movementexemplified by the writings of William Morris. This in turn formed the basis for Art Nouveau in the UK, exemplified by the work of Charles Rennie Mackintosh, and influenced theVienna Secession. On the Continent, the theories of Viollet-le-Duc and Gottfried Semper provided the springboard for enormous vitality of thought dedicated to architectural innovation and the renovation of the notion of style. Semper in particular developed an international following, in Germany, England, Switzerland, Austria, Bohemia, France, Italyand the United States.

The generation born during the middle-third of the 19th century was largely enthralled with the opportunities presented by Semper's combination of a breathtaking historical scope and a methodological granularity. In contrast to more recent, and thus "modern", thematically self-organized theoretical activities, this generation did not coalesce into a "movement." They did, however, seem to converge on Semper's use of the concept of *Realismus*, and they are thus labelled proponents of architectural realism. Among the most active Architectural Realists were: Georg Heuser, Rudolf Redtenbacher, Constantin Lipsius, Hans Auer, Paul Sédille, Lawrence Harvey, Otto Wagnerand Richard Streiter.

20th century

In 1889 Camillo Sitte published *Der Städtebau nach seinem künstlerischen Grundsätzen* (translated as *City Planning According to Artistic Principles*) which was not exactly a criticism of architectural form but an aesthetic criticism (inspired by medieval and Baroque town planning) of 19th-century urbanism. Mainly a theoretical work, it had an immediate impact on architecture, as the two disciplines of architecture and planning intertwined. Demand for it was so high that five editions appeared in German between 1889 and 1922 and a French translation came out in 1902.

(No English edition came out until 1945.) For Sitte, the most important issue was not the architectural shape or form of a building but the quality of the urban spaces that buildings collectively enclose, the whole being more than the sum of its parts. The Modern Movement rejected these thoughts and Le Corbusier energetically dismissed the work. Nevertheless, Sitte's work was revisited by post-modern architects and theorists from the 1970s, especially following its republication in 1986 by Rizzoli, in an edition edited by Collins and Collins (now published by Dover). The book is often cited anachronistically today as a vehicle for the criticism of the Modern Movement.

Also on the topic of artistic notions with regard to urbanism was Louis Sullivan's *The Tall Office Building Artistically Considered* of 1896. In this essay,

Sullivan penned his famous alliterative adage "form ever follows function"; a phrase that was to be later adopted as a central tenet of Modern architectural theory. While later architects adopted the abbreviated phrase "form follows function" as a polemic in service of functionalist doctrine, Sullivan wrote of function with regard to biological functions of the natural order. Another influential planning theorist of this time was Ebenezer Howard, who founded the garden city movement. This movement aimed to form communities with architecture in the Arts and Crafts style at Letchworth and Welwyn Garden City and popularised the style as domestic architecture.

In Vienna, the idea of a radically new modern architecture had many theorists and proponents. An early use of the term *modern architecture* in print occurred in the title of a book by Otto Wagner, who gave examples of his own work representative of the Vienna Secession with art nouveau illustrations, and didactic teachings to his students. Soon thereafter, Adolf Loos wrote *Ornament and Crime*, and while his own style is usually seen in the context of the Jugendstil, his demand for "the elimination of ornament" joined the slogan "form follows function" as a principle of the architectural so-called Modern Movement that came to dominate the mid-20th century.

Walter Gropius, Ludwig Mies van der Rohe and Le Corbusier provided the theoretical basis for the International Style with aims of using industrialised architecture to reshape society. Frank Lloyd Wright, while modern in rejecting historic revivalism, was idiosyncratic in his theory, which he conveyed in copious writing. Wright did not subscribe to the tenets of the International Style, but evolved what he hoped would be an American, in contrast to a European, progressive course. Wright's style, however, was highly personal, involving his particular views of man and nature. Wright was more poetic and firmly maintained the 19th-century view of the creative artist as unique genius. This limited the relevance of his theoretical propositions. Towards the end of the century postmodern architecture reacted against the austerity of High Modern (International Style) principles, viewed as narrowly normative and doctrinaire.

Contemporary

In contemporary architectural discourse theory has become more concerned with its position within culture generally, and thought in particular. This is why university courses on architecture theory may often spend just as much time discussing philosophy and cultural studies as buildings, and why advanced postgraduate research and doctoral dissertations focus on philosophical topics in connection with architectural humanities. Some architectural theorists aim at discussing philosophical themes, or engage in direct dialogues with philosophers, as in the case of Peter Eisenman's interest in Derrida's thought, or Christian Norberg-Schulz's interest in the works of Heidegger, in addition to an interest in Gaston Bachelard's *Poetics of Space* or texts by Gilles Deleuze.

This has also been the case with educators in academia like Dalibor Vesely or Alberto-Perez Gomez, and in more recent years this philosophical orientation has been reinforced through the research of a new generation of younger theorists, such as the philosopher-architectNader El-Bizri (specialist in history of classical optics and in phenomenology) or the academic-architect Adam Sharr. Similarly, we can refer to contemporary architects who are interested in phenomenology, like Steven Holl, Peter Zumthor and Juhani Pallasmaa, and who are referred to as "phenomenologists".

The notion that theory entailed critique also stemmed from post-structural literary studies. This, however, pushed architecture towards the notion of avant-gardism for its own sake - in many ways repeating the 19th-century *art for art's sake* outlook. Since 2000 this has materialised in architecture through concerns with the rapid rise of urbanism and globalization, but also a pragmatic understanding that the city can no longer be a homogeneous totality. Interests in fragmentation and architecture as transient objects further affected such thinking (*e.g.* the concern for employing high technology) but also related to general concerns such as ecology, mass media, and economism.

In the past decade, there has been a resurgence of proto-Modern "organic design" theories, but in a supposedly more scientific setting. Several currents and design methodologies are being developed simultaneously, some of which reinforce each other whereas others work in opposition. One of these trends is Biomimicry, which is the process of examining nature, its models, systems, processes, and elements to emulate or take inspiration from in order to solve human problems. Architects also design organic-looking buildings in the belief that by copying nature, Organic architecture reaches a more attractive or (more frequently) more efficient form.

Another trend is the exploration of those computational techniques that are influenced by algorithms relevant to biological processes and sometimes referred to as Digital morphogenesis. Trying to utilize Computational creativity in architecture, Genetic algorithms developed in computer science are applied to evolve designs on a computer, and some of these are proposed and built as actual structures. There exists, however, a controversy as to whether all such evolved designs through Design computing are truly appropriate for buildings or are instruments of self-deception dependent on the misapplication of biological analogies and metaphors, hence seeing such developments as being still ambiguous in their epistemic implications and theoretical foundations.

The new discipline of biophilia developed by E. O. Wilson suggests the advantages of forms inspired by biological structures, but in a more profound way than simple mimicry. Wilson's original idea is extended by Stephen R. Kellert in the Biophilia hypothesis, and applied to architectural design in the book "Biophilic Design". Mathematical features of biological forms such as fractals, scale-invariance, very sophisticated notions of symmetry, self-similarity, and complex hierarchy are proposed as essential tools for designing architectural forms.

Trying to understand the complex interaction between humans and their environment gained from human-computer interaction, mobile robotics, and artificial intelligence leads to ideas in intelligence-based design.

All these developments, though minuscule and highly localised in terms of total architectural output, give some observers (notably Harry Francis Mallgrave of the College of Architecture at the Illinois Institute of Technology) evidence for claiming that we are witnessing the birth of an entirely new type of architectural theory bearing little resemblance to the dominant school of architectural theory based on linguistic analysis, philosophy, post-structuralism, or cultural theory. It is too early, however, to say whether any of these explorations will have widespread or lasting impact.

ARCHITECTURAL TECHNOLOGY

Architectural Technology is a discipline related to the design of buildings. It is a new discipline which emerged from the practice ofarchitecture and building engineering. It was created as new technologies generated new design and construction methods. Architectural technology is related to the different elements of a building and their interactions.

In his published research, Stephen Emmitt explains that "*The relationship between building technology and design can be traced back to the Enlightenment and the industrial revolution, period when advances in technology and science were seen as the way forward, and times of solid faith in progress [...] As technologies multiply in number and complexity the building profession started to fragment. Increases in building activities brought about social and cultural changes*". We can assume that the practice of architectural technology and the practice of architecture were first dissociated during the period described by Stephen Emitt.

Fig. Façade detail - Center Pompidou in Paris

ACADEMIC DEFINITIONS

Universities define the discipline as: *"The technical design and expertise used in the application and integration of construction technologies in the building design process."* or as *"The ability to analyse, synthesise and evaluate building design factors in order to produce efficient and effective technical design solutions which satisfy performance, production and procurement criteria."*

PRACTICE OF ARCHITECTURAL TECHNOLOGY

In some countries the discipline is practiced by architectural technologists, while in other countries it is practiced by architects and engineers. Communication, planning, design, and technology are the core components of the profession, while environmental issues also became predominant.

Fig. César Pelli's Ratner Athletic Centeruses cables, counterweights and mastsas load-bearing devices.

The main subjects examined by architectural technology can be grouped as, but not limited to, the following disciplines:

- structural technologies;
- Architectural design
- building materials and methods;
- computer-aided design;
- environmental technologies applied to architecture;
- building management technologies;
- Building regulations and building standards.

Architectural technology cannot be limited to the subjects above. The practice of this disciple also requires knowledge of:

- Planning
- Project Management
- Contract Management
- Building control
- Building science

SUSTAINABLE ARCHITECTURE

Sustainable architecture is architecture that seeks to minimize the negative environmental impact of buildings by efficiency and moderation in the use of materials, energy, and development space. Sustainable architecture uses a conscious approach to energy and ecological conservation in the design of the built environment.

Fig. Energy-plus-houses at Freiburg-Vauban in Germany

The idea of sustainability, or ecological design, is to ensure that our actions and decisions today do not inhibit the opportunities of future generations.

SUSTAINABLE ENERGY USE

Energy efficiency over the entire life cycle of a building is the single most important goal of sustainable architecture. Architects use many different techniques to reduce the energy needs of buildings and increase their ability to capture or generate their own energy.

Heating, ventilation and cooling system efficiency

The most important and cost-effective element of an efficient heating, ventilating, and air conditioning (HVAC) system is a well-insulated building. A more efficient building requires less heat generating or dissipating power, but may require more ventilation capacity to expelpolluted indoor air.

Significant amounts of energy are flushed out of buildings in the water, air and compost streams. Off the shelf, on-site energy recycling technologies can effectively recapture energy from waste hot water and stale air and transfer that energy into incoming fresh cold water or fresh air. Recapture of energy for uses other than gardening from compost leaving buildings requires centralized anaerobic digesters.

HVAC systems are powered by motors. Copper, versus other metal conductors, helps to improve the electrical energy efficiencies of motors, thereby enhancing the sustainability of electrical building components.

Site and building orientation have some major effects on a building's HVAC efficiency.

Passive solar building design allows buildings to harness the energy of the sun efficiently without the use of any active solar mechanisms such as photovoltaic cells or solar hot water panels. Typically passive solar building designs incorporate materials with high thermal massthat retain heat effectively and strong insulation that works to prevent heat escape. Low energy designs also requires the use of solar shading, by means of awnings, blinds or shutters, to relieve the solar heat gain in summer and to reduce the need for artificial cooling. In addition, low energy buildings typically have a very low surface area to volume ratio to minimize heat loss.

This means that sprawling multi-winged building designs (often thought to look more "organic") are often avoided in Favour of more centralized structures. Traditional cold climate buildings such as American colonial saltbox designs provide a good historical model for centralized heat efficiency in a small-scale building.

Windows are placed to maximize the input of heat-creating light while minimizing the loss of heat through glass, a poor insulator. In thenorthern hemisphere this usually involves installing a large number of south-facing windows to collect direct sun and severely restricting the number of north-facing windows.

Certain window types, such as double or triple glazed insulated windows with gas filled spaces andlow emissivity (low-E) coatings, provide much better insulation than single-pane glass windows. Preventing excess solar gain by means of solar shading devices in the summer months is important to reduce cooling needs. Deciduous trees are often planted in front of windows to block excessive sun in summer with their leaves but allow light through in winter when their leaves fall off.

Louvers or light shelves are installed to allow the sunlight in during the winter (when the sun is lower in the sky) and keep it out in the summer (when the sun is high in the sky). Coniferous or evergreen plants are often planted to the north of buildings to shield against cold north winds.In colder climates, heating systems are a primary focus for sustainable architecture because they are typically one of the largest single energy drains in buildings.

In warmer climates where cooling is a primary concern, passive solar designs can also be very effective. Masonry building materials withhigh thermal mass are very valuable for retaining the cool temperatures of night throughout the day.

In addition builders often opt for sprawling single story structures in order to maximize surface area and heat loss. Buildings are often designed to capture and channel existing winds, particularly the especially cool winds coming from nearby bodies of water.

Many of these valuable strategies are employed in some way by the traditional architecture of warm regions, such as south-western mission buildings.

In climates with four seasons, an integrated energy system will increase in efficiency: when the building is well insulated, when it is sited to work with the forces of nature, when heat is recaptured (to be used immediately or stored), when the heat plant relying on fossil fuels or electricity is greater than 100per cent efficient, and when renewable energy is used.

RENEWABLE ENERGY GENERATION

Fig. BedZED (Beddington Zero Energy Development), the UK's largest and first carbon-neutral eco-community: the distinctive roofscape with solar panels and passive ventilation chimneys

Solar panels

Active solar devices such as photovoltaic solar panels help to provide sustainable electricity for any use. Electrical output of a solar panel is dependent on orientation, efficiency, latitude, and climate—solar gain varies even at the same latitude. Typical efficiencies for commercially available PV panels range from 4per cent to 28per cent. The low efficiency of certain photovoltaic panels can significantly affect the payback period of their installation. This low efficiency does not mean that solar panels are not a viable energy alternative. In Germany for example, Solar Panels are commonly installed in residential home construction.

Roofs are often angled toward the sun to allow photovoltaic panels to collect at maximum efficiency. In the northern hemisphere, a true-south facing orientation maximizes yield for solar panels. If true-south is not possible, solar panels can produce adequate energy if aligned within 30° of south. However, at higher latitudes, winter energy yield will be significantly reduced for non-south orientation.

To maximize efficiency in winter, the collector can be angled above horizontal Latitude +15°. To maximize efficiency in summer, the angle should be Latitude -15°. However, for an annual maximum production, the angle of the panel above horizontal should be equal to its latitude.

Wind turbines

The use of undersized wind turbines in energy production in sustainable structures requires the consideration of many factors. In considering costs, small wind systems are generally more expensive than larger wind turbines relative to the amount of energy they produce. For small wind turbines, maintenance costs can be a deciding factor at sites with marginal wind-harnessing capabilities. At low-wind sites, maintenance can consume much of a small wind turbine's revenue. Wind turbines begin operating when winds reach 8 mph, achieve energy production capacity at speeds of 32-37 mph, and shut off to avoid damage at speeds exceeding 55 mph. The energy potential of a wind turbine is proportional to the square of the length of its blades and to the cube of the speed at which its blades spin.

Though wind turbines are available that can supplement power for a single building, because of these factors, the efficiency of the wind turbine depends much upon the wind conditions at the building site. For these reasons, for wind turbines to be at all efficient, they must be installed at locations that are known to receive a constant amount of wind (with average wind speeds of more than 15 mph), rather than locations that receive wind sporadically.

A small wind turbine can be installed on a roof. Installation issues then include the strength of the roof, vibration, and the turbulence caused by the roof ledge. Small-scale rooftop wind turbines have been known to be able to generate power from 10per cent to up to 25per cent of the electricity required of a regular domestic household dwelling.

Turbines for residential scale use are usually between 7 feet (2 m) to 25 feet (8 m) in diameter and produce electricity at a rate of 900 watts to 10,000 watts at their tested wind speed. Building integrated wind turbine performance can be enhanced with the addition of an aerofoil wing on top of a roof mounted turbine.

Solar water heating

Solar water heaters, also called solar domestic hot water systems, can be a cost-effective way to generate hot water for a home. They can be used in any climate, and the fuel they use—sunshine—is free.

There are two types of solar water systems- active and passive. An active solar collector system can produce about 80 to 100 gallons of hot water per day. A passive system will have a lower capacity.

There are also two types of circulation, direct circulation systems and indirect circulation systems. Direct circulation systems loop the domestic water through the panels. They should not be used in climates with temperatures below freezing. Indirect circulation loops glycol or some other fluid through the solar panels and uses a heat exchanger to heat up the domestic water.

The two most common types of collector panels are Flat-Plate and Evacuated-tube. The two work similarly except that evacuated tubes do not convectively lose heat, which greatly improves their efficiency (5per cent-25per cent more efficient). With these higher efficiencies, Evacuated-tube solar collectors can also produce higher-temperature space heating, and even higher temperatures for absorption cooling systems.

Electric-resistance water heaters that are common in homes today have an electrical demand around 4500 kW·h/year. With the use of solar collectors, the energy use is cut in half. The up-front cost of installing solar collectors is high, but with the annual energy savings, payback periods are relatively short.

Heat pumps

Air-source heat pumps (ASHP) can be thought of as reversible air conditioners. Like an air conditioner, an ASHP can take heat from a relatively cool space (*e.g.* a house at 70°F) and dump it into a hot place (*e.g.* outside at 85°F). However, unlike an air conditioner, the condenser and evaporator of an ASHP can switch roles and absorb heat from the cool outside air and dump it into a warm house.

Air-source heat pumps are inexpensive relative to other heat pump systems. However, the efficiency of air-source heat pumps decline when the outdoor temperature is very cold or very hot; therefore, they are only really applicable in temperate climates.

For areas not located in temperate climates, ground-source (or geothermal) heat pumps provide an efficient alternative. The difference between the two heat pumps is that the ground-source has one of its heat exchangers placed underground—usually in a horizontal or vertical arrangement. Ground-source takes advantage of the relatively constant, mild temperatures underground, which means their efficiencies can be much greater than that of an air-source heat pump. The in-ground heat exchanger generally needs a considerable amount of area. Designers have placed them in an open area next to the building or underneath a parking lot.

Energy Star ground-source heat pumps can be 40per cent to 60per cent more efficient than their air-source counterparts. They are also quieter and can also be applied to other functions like domestic hot water heating.

In terms of initial cost, the ground-source heat pump system costs about twice as much as a standard air-source heat pump to be installed. However, the up-front costs can be more than offset by the decrease in energy costs. The reduction in energy costs is especially apparent in areas with typically hot summers and cold winters.

Other types of heat pumps are water-source and air-earth. If the building is located near a body of water, the pond or lake could be used as a heat source or sink. Air-earth heat pumps circulate the building's air through underground ducts. With higher fan power requirements and inefficient heat transfer, Air-earth heat pumps are generally not practical for major construction.

SUSTAINABLE BUILDING MATERIALS

Some examples of sustainable building materials include recycled denim or blown-in Fibre glass insulation, sustainably harvested wood, Trass, Linoleum, sheep wool,concrete (high and ultra high performance roman self-healing concrete), panels made from paper flakes, baked earth, rammed earth, clay, vermiculite, flax linnen, sisal, seegrass, expanded clay grains, coconut, wood fibre plates, calcium sand stone, locally obtained stone and rock, and bamboo, which is one of the strongest and fastest growingwoody plants, and non-toxic low-VOC glues and paints.

RECYCLED MATERIALS

Sustainable architecture often incorporates the use of recycled or second hand materials, such as reclaimed lumber and recycled copper. The reduction in use of new materials creates a corresponding reduction in embodied energy (energy used in the production of materials). Often sustainable architects attempt to retrofit old structures to serve new needs in order to avoid unnecessary development. Architectural salvage and reclaimed materials are used when appropriate. When older buildings are demolished, frequently any good wood is reclaimed, renewed, and sold as flooring. Any good dimension stone is similarly reclaimed. Many other parts are reused as well, such as doors, windows, mantels, and hardware, thus reducing the consumption of new goods. When new materials are employed, green designers look for materials that are rapidly replenished, such as bamboo, which can be harvested for commercial use after only 6 years of growth, sorghum or wheat straw, both of which are waste material that can be pressed into panels, or cork oak, in which only the outer bark is removed for use, thus preserving the tree.

When possible, building materials may be gleaned from the site itself; for example, if a new structure is being constructed in a wooded area, wood from the trees which were cut to make room for the building would be re-used as part of the building itself.

LOWER VOLATILE ORGANIC COMPOUNDS

Low-impact building materials are used wherever feasible: for example, insulation may be made from low VOC (volatile organic compound)-emitting materials such as recycled denim or cellulose insulation, rather than the building insulation materials that may contain carcinogenic or toxic materials such as formaldehyde. To discourage insect damage, these alternate insulation materials may be treated with boric acid.

Organic or milk-based paints may be used. However, a common fallacy is that "green" materials are always better for the health of occupants or the environment. Many harmful substances (including formaldehyde, arsenic, and asbestos) are naturally occurring and are not without their histories of use with the best of intentions. A study of emissions from materials by the State of California has shown that there are some green materials that have substantial emissions whereas some more "traditional" materials actually were lower emitters. Thus, the subject of emissions must be carefully investigated before concluding that natural materials are always the healthiest alternatives for occupants and for the Earth.

Volatile organic compounds (VOC) can be found in any indoor environment coming from a variety of different sources. VOCs have a high vapor pressure and low water solubility, and are suspected of causing sick building syndrome type symptoms. This is because many VOCs have been known to cause sensory irritation and central nervous system symptoms characteristic to sick building syndrome, indoor concentrations of VOCs are higher than in the outdoor atmosphere, and when there are many VOCs present, they can cause additive and multiplicative effects.

Green products are usually considered to contain fewer VOCs and be better for human and environmental health. A case study conducted by the Department of Civil, Architectural, and Environmental Engineering at the University of Miami that compared three green products and their non-green counterparts found that even though both the green products and the non-green counterparts both emitted levels of VOCs, the amount and intensity of the VOCs emitted from the green products were much safer and comfortable for human exposure.

MATERIALS SUSTAINABILITY STANDARDS

Despite the importance of materials to overall building sustainability, quantifying and evaluating the sustainability of building materials has proven difficult. There is little coherence in the measurement and assessment of materials sustainability attributes, resulting in a landscape today that is littered with hundreds of competing, inconsistent and often imprecise eco-labels, standards and certifications.

This discord has led both to confusion among consumers and commercial purchasers and to the incorporation of inconsistent sustainability criteria in larger building certification Programmes such as LEED. Various proposals have been made regarding rationalization of the standardization landscape for sustainable building materials.

WASTE MANAGEMENT

Waste takes the form of spent or useless materials generated from households and businesses, construction and demolition processes, and manufacturing and agricultural industries. These materials are loosely categorized as municipal solid waste, construction and demolition (CandD) debris, and industrial or agricultural by-products. Sustainable architecture focuses on the on-site use of waste management, incorporating things such as grey water systems for use on garden beds, and composting toilets to reduce sewage.

These methods, when combined with on-site food waste composting and off-site recycling, can reduce a house's waste to a small amount of packaging waste.This is the new techniques of sustainable architecture.

BUILDING PLACEMENT

One central and often ignored aspect of sustainable architecture is building placement. Although the ideal environmental home or office structure is often envisioned as an isolated place, this kind of placement is usually detrimental to the environment. First, such structures often serve as the unknowing frontlines of suburban sprawl. Second, they usually increase the energy consumption required for transportation and lead to unnecessary auto emissions. Ideally, most building should avoid suburban sprawl in Favour of the kind of light urban development articulated by the New Urbanist movement. Careful mixed use zoning can make commercial, residential, and light industrial areas more accessible for those traveling by foot, bicycle, or public transit, as proposed in the Principles of Intelligent Urbanism. The study of Permaculture, in its holistic application, can also greatly help in proper building placement that minimizes energy consumption and works with the surroundings rather than against them, especially in rural and forested zones.

SUSTAINABLE BUILDING CONSULTING

A sustainable building consultant may be engaged early in the design process, to forecast the sustainability implications of building materials, orientation, glazing and other physical factors, so as to identify a sustainable approach that meets the specific requirements of a project.

Norms and standards have been formalized by performance-based rating systems *e.g.* LEED and Energy Star for homes. They define benchmarks to be met and providemetrics and testing to meet those benchmarks. It is up to the parties involved in the project to determine the best approach to meet those standards.

CHANGING PEDAGOGIES

Critics of the reductionism of modernism often noted the abandonment of the teaching of architectural history as a causal factor. The fact that a number of the major players in the shift away from modernism were trained at Princeton University's School of Architecture, where recourse to history continued to be a part of design training in the 1940s and 1950s, was significant. The increasing rise of interest in history had a profound impact on architectural education. History courses became more typical and regularized. With the demand for professors knowledgeable in the history of architecture, several PhD Programmes in schools of architecture arose in order to differentiate themselves from art history PhD Programmes, where architectural historians had previously trained. In the US, MIT and Cornell were the first, created in the mid-1970s, followed by Columbia, Berkeley, andPrinceton. Among the founders of new architectural history Programmes were Bruno Zevi at the Institute for the History of Architecture in Venice, Stanford Anderson and Henry Millon at

MIT, Alexander Tzonis at the Architectural Association, Anthony Vidler at Princeton, Manfredo Tafuri at the University of Venice, Kenneth Frampton at Columbia University, and Werner Oechslin and Kurt Forster at ETH Zürich.

The term "sustainability" in relation to architecture has so far been mostly considered through the lens of building technology and its transformations. Going beyond the technical sphere of "green" design, invention and expertise, some scholars are starting to position architecture within a much broader cultural framework of the human interrelationship with nature. Adopting this framework allows tracing a rich history of cultural debates about our relationship to nature and the environment, from the point of view of different historical and geographical contexts.

SUSTAINABLE URBANISM AND ARCHITECTURE

Concurrently, the recent movements of New Urbanism and New Classical Architecture promote a sustainable approach towards construction, that appreciates and developssmart growth, architectural tradition and classical design. This in contrast to modernist and globally uniform architecture, as well as leaning against solitary housing estatesand suburban sprawl. Both trends started in the 1980s. The Driehaus Architecture Prize is an award that recognizes efforts in New Urbanism and New Classical Architecture, and is endowed with a prize money twice as high as that of the modernist Pritzker Prize.

3

Structural Engineering

Structural engineers examine, design, plan, and research structural components and structural systems to achieve design goals and ensure the safety and comfort of users or occupants. Their work takes account mainly of safety, technical, economic and environmental concerns, but they may also consider aesthetic and social factors.

Structural engineering is usually considered a specialty discipline within civil engineering, but it can also be studied in its own right. In the US, most practicing structural engineers are currently licensed as civil engineers, but the situation varies from state to state.

In UK, most structural engineers in the building industry are members of the Institution of Structural Engineers rather than the Institution of Civil Engineers.

Typical structures designed by a structural engineer include buildings, towers, stadium and bridges. Other structures such as oil rigs, space satellites, aircraft and ships may also be designed by a structural engineer. Most structural engineers are employed in the construction industry, however there are also structural engineers in the aerospace, automobile and shipbuilding industries. In the construction industry, they work closely with architects, civil engineers, mechanical engineers, electrical engineers, quantity surveyors, and construction managers.

Structural engineers make sure that buildings and bridges are built to be strong enough and stable enough to resist all appropriate structural loads that includes:

- Gravity
- Wind
- Snow
- Rain
- Earthquake
- Earth pressure
- Temperature
- Traffic

In order to prevent or reduce loss of life or injury. They also design structures to be stiff enough to not deflect or vibrate beyond acceptable limits. Human comfort is an issue that is regularly considered in the limits. Fatigue is also an important consideration for bridges and for aircraft design or for other structures which experience a large number of stress cycles over their lifetimes. Consideration is also given to durability of materials against possible deterioration which may impair performance over the design lifetime.

The education of structural engineers is usually through a civil engineering bachelor's degree, and often a master's degree specializing in structural engineering.

The fundamental core subjects for structural engineering are strength of materials or solid mechanics, Structural Analysis -Static and Dynamic, material science, numerical analysis and conceptual structural design. Reinforced concrete, composite structure, timber, masonry and structural steel designs are the general structural design courses that will be introduced in the next level of the education of structural engineering.

The structural analysis courses which include structural mechanics, structural dynamics and structural failure analyses are designed to build up the fundamental analyses skills and theories for structural engineering students. At the senior year level or in graduate Programmes, prestressed concrete design, space frame design for building and aircraft, bridge engineering, civil and aerospace structure rehabilitation and other advanced structural engineering specializations are usually introduced.

Recently in the United States, there have been discussions in the structural engineering community about the knowledge base of structural engineering graduates. Some have called for a master's degree to be the minimum standard for professional licensing as a civil engineer. Many students who later become structural engineers in areas of civil, mechanical, or aerospace engineering, emphasis on structural engineering. Architectural engineering also offer structural emphases, and are often in combined academic departments along with civil engineering.

Structural engineering is a field of engineering that deals with the analysis and design of structures that support or resist loads. Structural engineering is usually considered a specialty within civil engineering, but it can also be studied in its own right.

Structural engineers are most commonly involved in the design of buildings and large non-building structures but they can also be involved in the design of machinery, medical equipment, vehicles or any item where structural integrity affects the item's function or safety. Structural engineers must ensure their designs satisfy given design criteria, predicated on safety in relation with structures that should not collapse without due warning or serviceability and performance.

Structural engineering theory is based upon physical laws and empirical knowledge of the structural performance of different materials and geometries. Structural engineering design utilizes a number of simple structural elements to build complex structural systems. Structural engineers are responsible for making creative and efficient use of funds, structural elements and materials to achieve these goals.

Fig. Structural engineering deals with the making of complex systems like theInternational Space Station, here seen from the departing Space Shuttle*Atlantis*.

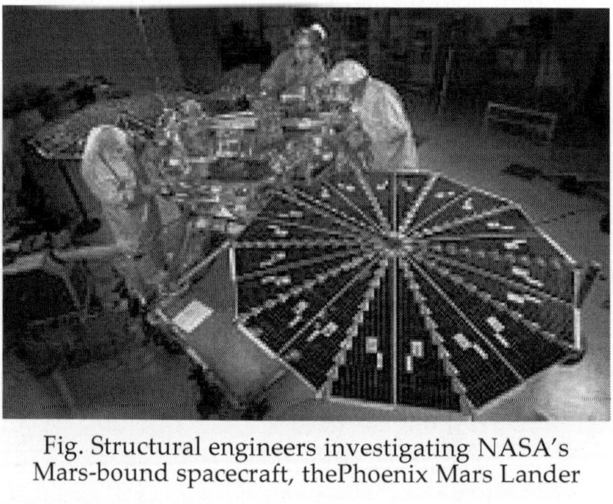

Fig. Structural engineers investigating NASA's Mars-bound spacecraft, thePhoenix Mars Lander

Fig. The Eiffel Tower is a historical achievement of structural engineering.

Structural engineering is a field of engineering dealing with the analysis and design of structures that support or resist loads.

Structural engineers are most commonly involved in the design of buildings and large Non-building structures but they can also be involved in the design of machinery, medical equipment, vehicles or any item where structural integrity affects the item's function or safety.

Structural engineers must ensure their designs satisfy given design criteria, predicated on safety (*e.g.* structures must not collapse without due warning) or serviceability and performance (*e.g.* building sway must not cause discomfort to the occupants).

Structural engineering theory is based upon applied physical laws and empirical knowledge of the structural performance of different materials and geometries.

Structural engineering design utilizes a number of simple structural elements to build complex structural systems. Structural engineers are responsible for making creative and efficient use of funds, structural elements and materials to achieve these goals.

WHAT IS STRUCTURAL ENGINEERING?

Structural engineering involves the analysis and design of structures such as buildings, bridges, towers, marine structures, dams, tunnels, retaining walls and other infrastructure.

Structural engineering underpins and sustains the built environment, where structures must be safe, serviceable, durable, aesthetically pleasing and economical.

Structural engineering applies maths and physics to traditional construction materials such as concrete, stone, steel, timber and glass and innovative engineering materials, including aluminium, polymers and carbon fibre.

Structural design

Structural design determines the type of structure that is suitable for a particular purpose, the materials to be used, the loads and other actions that the structure must sustain, and the arrangement, layout and dimensions of its various components.

Structural design involves detailed calculations to ensure that:

- the structure is stable
- all parts have adequate strength to resist the design loads
- the structure as a whole will remain serviceable throughout its design life and able to perform its intended function.

Finally, structural design involves the careful preparation of drawings that will communicate the engineering design to the contractors who will build the structure.

Structural analysis

Structural analysis is an integral part of structural design. It involves the calculation of the response of the structure to the design loads and imposed deformations that it will be required to resist during its lifetime. In structural engineering, 'deformation' refers to when an object is changed temporarily or permanently due to applied force. These calculations allow structural engineers to select the right materials for the structure, and to ensure that it will be suitable for the purpose for which it is being built.

STRUCTURAL ENGINEER (PROFESSIONAL)

Structural engineers are responsible for engineering design and analysis. Entry-level structural engineers may design the individual structural elements of a structure, for example the beams, columns, and floors of a building. More experienced engineers may be responsible for the structural design and integrity of an entire system, such as a building.

Structural engineers often specialise in particular fields, such as bridge engineering, building engineering, pipeline engineering, industrial structures, or special mechanical structures such as vehicles, ships or aircraft.

Structural engineering has existed since humans first started to construct their own structures.

It became a more defined and formalised profession with the emergence of the architecture profession as distinct from the engineering profession during the industrial revolution in the late 19th century. Until then, the architect and thestructural engineer were usually one and the same - the master builder. Only with the development of specialised knowledge of structural theories that emerged during the 19th and early 20th centuries did the professional structural engineer come into existence.

The role of a structural engineer today involves a significant understanding of both static and dynamic loading, and the structures that are available to resist them.

The complexity of modern structures often requires a great deal of creativity from the engineer in order to ensure the structures support and resist the loads they are subjected to.

A structural engineer will typically have a four or five year undergraduate degree, followed by a minimum of three years of professional practice before being considered fully qualified. Structural engineers are licensed or accredited by different learned societies and regulatory bodies around the world (for example, the Institution of Structural Engineers in the UK).

Depending on the degree course they have studied and/or the jurisdiction they are seeking licensure in, they may be accredited (or licensed) as just structural engineers, or as civil engineers, or as both civil and structural engineers. Another international organisation is IABSE(Internation Association for Bridge and Structural Engineering). The aim of that association is to exchange knowledge and to advance the practice of structural engineering worldwide in the service of the profession and society.

HISTORY OF STRUCTURAL ENGINEERING

Fig. Pont du Gard, France, a Roman era aqueduct circa 19 BC.

Structural engineering dates back to 2700 B.C.E. when the step pyramid for Pharaoh Djoser was built byImhotep, the first engineer in history known by name. Pyramids were the most common major structures built by ancient civilizations because the structural form of a pyramid is inherently stable and can be almost infinitely scaled (as opposed to most other structural forms, which cannot be linearly increased in size in proportion to increased loads).

However, it is important to note that the structural stability of the pyramid is not primarily a result of its shape. The integrity of the pyramid is intact as long as each of the stones is able to support the weight of the stone above it. The limestone blocks were taken from a quarry near the build site. Since the compressive strength of limestone is anywhere from 30 to 250 MPa (MPa = Pa * 10 ^ 6), the blocks will not fail under compression. Therefore the structural strength of the pyramid stems from the material properties of the stones from which it was built rather than the pyramid's geometry.

Throughout ancient and medieval history most architectural design and construction was carried out byartisans, such as stone masons and carpenters, rising to the role of master builder. No theory of structures existed, and understanding of how structures stood up was extremely limited, and based almost entirely on empirical evidence of 'what had worked before'. Knowledge was retained by guilds and seldom supplanted by advances. Structures were repetitive, and increases in scale were incremental.

No record exists of the first calculations of the strength of structural members or the behaviour of structural material, but the profession of structural engineer only really took shape with the Industrial Revolution and the re-invention of concrete. The physical sciences underlying structural engineering began to be understood in the Renaissance and have since developed into computer-based applications pioneered in the 1970s.

Timeline

- 1452–1519 Leonardo da Vinci made many contributions
- 1638: Galileo Galilei published the book "Two New Sciences" in which he examined the failure of simple structures
- 1660: Hooke's law by Robert Hooke
- 1687: Isaac Newton published "Philosophiae Naturalis Principia Mathematica" which contains the Newton's laws of motion
- 1750: Euler–Bernoulli beam equation
- 1700–1782: Daniel Bernoulli introduced the principle of virtual work
- 1707–1783: Leonhard Euler developed the theory of buckling of columns
- 1826: Claude-Louis Navier published a treatise on the elastic behaviors of structures
- 1873: Carlo Alberto Castigliano presented his dissertation "Intorno ai sistemi elastici", which contains his theorem for computing displacement as partial derivative of the strain energy. This theorem includes the method of *least work* as a special case
- 1874: Otto Mohr formalised the idea of a statically indeterminate structure.
- 1922: Timoshenko corrects the Euler-Bernoulli beam equation
- 1936: Hardy Cross' publication of the moment distribution method which was later recognised as a form of the relaxation method applicable to the problem of flow in pipe-network
- 1941: Alexander Hrennikoff submitted his D.Sc thesis in MIT on the discretisation of plane elasticity problems using a lattice framework
- 1942: R. Courant divided a domain into finite subregions
- 1956: J. Turner, R. W. Clough, H. C. Martin, and L. J. Topp's paper on the "Stiffness and Deflection of Complex Structures" introduces the name "finite-element method" and is widely recognised as the first comprehensive treatment of the method as it is known today

Structural failure

The history of structural engineering contains many collapses and failures. Sometimes this is due to obvious negligence, as in the case of the Pétionville school collapse, in which Rev. Fortin Augustin said that *"he constructed the building all by himself, saying he didn't need an engineer as he had good knowledge of construction"* following a partial collapse of the three-story schoolhouse that sent Neighbours fleeing. The final collapse killed 94 people, mostly children.

In other cases structural failures require careful study, and the results of these inquiries have resulted in improved practices and greater understanding of the science of structural engineering. Some such studies are the result offorensic engineering investigations where the original engineer seems to have done everything in accordance with the state of the profession and acceptable practice yet a failure still eventuated.

A famous case of structural knowledge and practice being advanced in this manner can be found in a series of failures involving box girders which collapsed in Australia during the 1970s.

SPECIALISATIONS

Building structures

Fig. Sydney Opera House, designed byOve Arup and Partners, with the architectJørn Utzon

Fig. Millennium Dome in London, UK, byBuro Happold and Richard Rogers

Fig. Burj Khalifa, in Dubai, the world's tallest building, shown under construction in 2007 (since completed)

Structural building engineering includes all structural engineering related to the design of buildings. It is the branch of structural engineering that is close to architecture.

Structural building engineering is primarily driven by the creative manipulation of materials and forms and the underlying mathematical and scientific ideas to achieve an end which fulfills its functional requirements and is structurally safe when subjected to all the loads it could reasonably be expected to experience.

This is subtly different from architectural design, which is driven by the creative manipulation of materials and forms, mass, space, volume, texture and light to achieve an end which is aesthetic, functional and often artistic.

The architect is usually the lead designer on buildings, with a structural engineer employed as a sub-consultant.

The degree to which each discipline actually leads the design depends heavily on the type of structure.

Many structures are structurally simple and led by architecture, such as multi-storey office buildings and housing, while other structures, such as tensile structures, shells and gridshells are heavily dependent on their form for their strength, and the engineer may have a more significant influence on the form, and hence much of the aesthetic, than the architect.

The structural design for a building must ensure that the building is able to stand up safely, able to function without excessive deflections or movements which may cause fatigue of structural elements, cracking or failure of fixtures, fittings or partitions, or discomfort for occupants. It must account for movements and forces due to temperature, creep, cracking and imposed loads. It must also ensure that the design is practically buildable within acceptable manufacturing tolerances of the materials.

It must allow the architecture to work, and the building services to fit within the building and function (air conditioning, ventilation, smoke extract, electrics, lighting etc.).

The structural design of a modern building can be extremely complex, and often requires a large team to complete.

Structural engineering specialties for buildings include:

- Earthquake engineering
- Façade engineering
- Fire engineering
- Roof engineering
- Tower engineering
- Wind engineering

Earthquake engineering structures

Earthquake engineering structures are those engineered to withstand earthquakes.

Fig. Earthquake-proof pyramid El Castillo, Chichen Itza

The main objectives of earthquake engineering are to understand the interaction of structures with the shaking ground, foresee the consequences of possible earthquakes, and design and construct the structures to perform during an earthquake.

Fig. Snapshot from shake-table video of testing base-isolated (right) and regular (left) building model

Earthquake-proof structures are not necessarily extremely strong like the El Castillo pyramid at Chichen Itza shown above. In fact, many structures considered strong may in fact be stiff, which can result in poor seismic performance.

One important tool of earthquake engineering is base isolation, which allows the base of a structure to move freely with the ground.

Civil engineering structures

Civil structural engineering includes all structural engineering related to the built environment. It includes:

- Bridges
- Dams
- Earthworks
- Foundations
- Offshore structures
- Pipelines
- Power stations

- Railways
- Retaining structures and walls
- Roads
- Tunnels
- Waterways
- Water and wastewater infrastructure

The structural engineer is the lead designer on these structures, and often the sole designer. In the design of structures such as these, structural safety is of paramount importance (in the UK, designs for dams, nuclear power stations and bridges must be signed off by a chartered engineer).

Civil engineering structures are often subjected to very extreme forces, such as large variations in temperature, dynamic loads such as waves or traffic, or high pressures from water or compressed gases

They are also often constructed in corrosive environments, such as at sea, in industrial facilities or below ground.

Mechanical structures

Fig. Mechanical Structures

Fig. Principles of structural engineering are applied to variety of mechanical (moveable) structures. The design of static structures assumes they always have the same geometry (in fact, so-called static structures can move significantly, and structural engineering design must take this into account where necessary), but the design of moveable or moving structures must account for fatigue, variation in the method in which load is resisted and significant deflections of structures.

The forces which parts of a machine are subjected to can vary significantly, and can do so at a great rate. The forces which a boat or aircraft are subjected to vary enormously and will do so thousands of times over the structure's lifetime. The structural design must ensure that such structures are able to endure such loading for their entire design life without failing.

These works can require mechanical structural engineering:

- Boilers and pressure vessels
- Coachworks and carriages
- Cranes
- Elevators
- Escalators
- Marine vessels and hulls

Aerospace structures

Fig. An Airbus A380, the world's largest passenger airliner

Fig. Design of missile needs in depth understanding of Structural Analysis

Aerospace structure types include launch vehicles, (Atlas, Delta, Titan), missiles (ALCM, Harpoon), Hypersonic vehicles (Space Shuttle), military aircraft (F-16, F-18) and commercial aircraft (Boeing 777, MD-11). Aerospace structures typically consist of thin plates with stiffeners for the external surfaces, bulkheads and frames to support the shape and fasteners such as welds, rivets, screws and bolts to hold the components together.

Nanoscale structures

A nanostructure is an object of intermediate size between molecular and microscopic (micrometer-sized) structures. In describing nanostructures it is necessary to differentiate between the number of dimensions on the nanoscale. Nanotextured surfaces have one dimension on the nanoscale, *i.e.*, only the thickness of the surface of an object is between 0.1 and 100 nm. Nanotubes have two dimensions on the nanoscale, *i.e.*, the diameter of the tube is between 0.1 and 100 nm; its length could be much greater.

Finally, spherical nanoparticles have three dimensions on the nanoscale, *i.e.*, the particle is between 0.1 and 100 nm in each spatial dimension. The terms nanoparticles and ultrafine particles (UFP) often are used synonymously although UFP can reach into the micrometre range. The term 'nanostructure' is often used when referring to magnetic technology.

Structural Engineering for Medical Science

Fig. Designing Medical Equipment needs in-depth understanding of Structural Engineering

Medical equipment (also known as armamentarium) is designed to aid in the diagnosis, monitoring or treatment of medical conditions. There are several basic types: Diagnostic equipment includes medical imaging machines, used to aid in diagnosis; equipment includes infusion pumps, medical lasers and LASIK surgical machines; Medical monitors allow medical staff to measure a patient's medical state. Monitors may measure patient vital signs and other parameters including ECG, EEG, blood pressure, and dissolved gases in the blood; Diagnostic Medical Equipment may also be used in the home for certain purposes, *e.g.* for the control of diabetes mellitus. A biomedical equipment technician (BMET) is a vital component of the Health care delivery system. Employed primarily by hospitals, BMETs are the people responsible for maintaining a facility's medical equipment.

STRUCTURAL ELEMENTS

Any structure is essentially made up of only a small number of different types of elements:

- Columns
- Beams
- Plates
- Arches
- Shells
- Catenaries

Many of these elements can be classified according to form (straight, plane/ curve) and dimensionality (one-dimensional/ two-dimensional):

	One-dimensional		Two-dimensional	
	straight	curve	plane	curve
(predominantly) bending	beam	continuous arch	plate, concrete slab	lamina, dome
(predominant) tensile stress	rope, tie	Catenary	shell	
(predominant) compression	Pier, column		Load-bearing wall	

COLUMNS

Columns are elements that carry only axial force - compression - or both axial force and bending (which is technically called a beam-column but practically, just a column). The design of a column must check the axial capacity of the element, and the buckling capacity.

The buckling capacity is the capacity of the element to withstand the propensity to buckle.

Its capacity depends upon its geometry, material, and the effective length of the column, which depends upon the restraint conditions at the top and bottom of the column. The effective length is K *l where l is the real length of the column.

The capacity of a column to carry axial load depends on the degree of bending it is subjected to, and vice versa. This is represented on an interaction chart and is a complex non-linear relationship.

BEAMS

Fig. Little Belt: a truss bridge in Denmark

A beam may be defined as an element in which one dimension is much greater than the other two and the applied loads are usually normal to the main axis of the element.

Beams and columns are called line elements and are often represented by simple lines in structural modeling.

- cantilevered (supported at one end only with a fixed connection)
- simply supported (supported vertically at each end; horizontally on only one to withstand friction, and able to rotate at the supports)
- fixed (supported at both ends by fixed connection; unable to rotate at the supports)
- continuous (supported by three or more supports)
- a combination of the above (ex. supported at one end and in the middle)

Beams are elements which carry pure bending only. Bending causes one part of the section of a beam (divided along its length) to go into compression and the other part into tension. The compression part must be designed to resist buckling and crushing, while the tension part must be able to adequately resist the tension.

TRUSSES

Fig. The McDonnell Planetarium by Gyo Obata in St Louis, Missouri, USA, a concrete shell structure

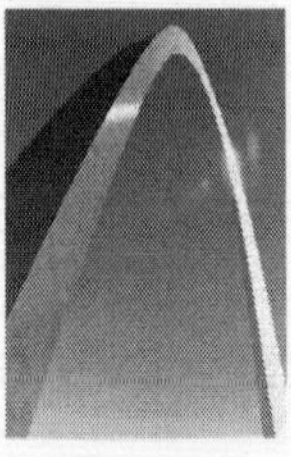

Fig. The 630 foot (192 m) high, stainless-clad (type 304) Gateway Arch in Saint Louis, Missouri

A truss is a structure comprising two types of structural elements; compression members and tension members (*i.e.* struts and ties). Most trusses use gusset plates to connect intersecting elements. Gusset plates are relatively flexible and minimize bending moments at the connections, thus allowing the truss members to carry primarily tension or compression.

Trusses are usually utilised in large-span structures, where it would be uneconomical to use solid beams.

PLATES

Plates carry bending in two directions. A concrete flat slab is an example of a plate. Plates are understood by using continuum mechanics, but due to the complexity involved they are most often designed using a codified empirical approach, or computer analysis.

They can also be designed with yield line theory, where an assumed collapse mechanism is analysed to give an upper bound on the collapse load. This technique is used in practice but because the method provides an upper-bound, *i.e.* an unsafe prediction of the collapse load, for poorly conceived collapse mechanisms great care is needed to ensure that the assumed collapse mechanism is realistic.

SHELLS

Shells derive their strength from their form, and carry forces in compression in two directions. A dome is an example of a shell. They can be designed by making a hanging-chain model, which will act as a catenary in pure tension, and inverting the form to achieve pure compression.

ARCHES

Arches carry forces in compression in one direction only, which is why it is appropriate to build arches out of masonry. They are designed by ensuring that the line of thrust of the force remains within the depth of the arch. It is mainly used to increase the bountifulness of any structure.

CATENARIES

Catenaries derive their strength from their form, and carry transverse forces in pure tension by deflecting (just as a tightrope will sag when someone walks on it). They are almost always cable or fabric structures. A fabric structure acts as a catenary in two directions.

STRUCTURAL ENGINEERING THEORY

Structural engineering depends upon a detailed knowledge of applied mechanics, materials science and applied mathematics to understand and predict how structures support and resist self-weight and imposed loads. To apply the knowledge successfully a structural engineer generally requires detailed knowledge of relevant empirical and theoretical design codes, the techniques of structural analysis, as well as some knowledge of the corrosion resistance of the materials and structures, especially when those structures are exposed to the external environment. Since the 1990s, specialist software has become available to aid in the design of structures, with the functionality to assist in the drawing, analyzing and designing of structures with maximum precision; examples include AutoCAD, StaadPro, ETABS, Prokon, Revit Structure etc. Such software may also take into consideration environmental loads, such as from earthquakes and winds.

MATERIALS

Structural engineering depends on the knowledge of materials and their properties, in order to understand how different materials support and resist loads.

Common structural materials are:

- Iron: Wrought iron, Cast iron
- Concrete: Reinforced concrete, Prestressed concrete
- Alloy: Steel, Stainless steel
- Masonry
- Timber: Hardwood, Softwood
- Aluminium
- Composite materials: Plywood
- Other structural materials:Adobe, Bamboo, Carbon fibre, Fibre reinforced plastic, Mudbrick, Roofing materials

HISTORY OF STRUCTURAL ENGINEERING

The history of structural engineering dates back to at least 2700 BC when the step pyramid for Pharaoh Djoser was built by Imhotep, the first engineer in history known by name. Pyramids were the most common major structures built by ancient civilizations because it is a structural form which is inherently stable and can be almost infinitely scaled (as opposed to most other structural forms, which cannot be linearly increased in size in proportion to increased loads).Another notable engineering feat from antiquity stiil in use today is the qanat water management system. Qanat technology developed in the time of the Medes, the predecessors of the Persian Empire (modern-day Iran which has the oldest and longest Qanat (older than 3000 years and longer than 71 km) that also spread to other cultures having had contact with the Persian.

Throughout ancient and medieval history most architectural design and construction was carried out by artisans, such as stone masons andcarpenters, rising to the role of master builder. No theory of structures existed and understanding of how structures stood up was extremely limited, and based almost entirely on empirical evidence of 'what had worked before'. Knowledge was retained by guilds and seldom supplanted by advances. Structures were repetitive, and increases in scale were incremental. No record exists of the first calculations of the strength of structural members or the behaviour of structural material, but the profession of structural engineer only really took shape with the industrial revolution and the re-invention of concrete. Thephysical sciences underlying structural engineering began to be understood in the Renaissance and have been developing ever since.

EARLY STRUCTURAL ENGINEERING DEVELOPMENTS

The recorded history of structural engineering starts with the ancient Egyptians. In the 27th century BC, Imhotep was the first structural engineer known by name and constructed the first known step pyramid in Egypt. In the 26th century BC, the Great Pyramid of Giza was constructed in Egypt. It remained the largest man-made structure for millennia and was considered an unsurpassed feat in architecture until the 19th century AD.

The understanding of the physical laws that underpin structural engineering in the Western world dates back to the 3rd century BC, when Archimedes published his work *On the Equilibrium of Planes* in two volumes, in which he sets out the *Law of the Lever*, stating:

Equal weights at equal distances are in equilibrium, and equal weights at unequal distances are not in equilibrium but incline towards the weight which is at the greater distance.

Archimedes used the principles derived to calculate the areas and centers of gravity of various geometric figures including triangles,paraboloids, and hemispheres. Archimedes's work on this and his work on calculus and geometry, together with Euclidean geometry, underpin much of the mathematics and understanding of structures in modern structural engineering.

Fig. Pont du Gard, France, a Roman era aqueduct circa 19 BC.

The ancient Romans made great bounds in structural engineering, pioneering large structures in masonry and concrete, many of which are still standing today. They include aqueducts, thermae, columns, lighthouses, defensive walls and harbours. Their methods are recorded by Vitruvius in his De Architectura written in 25 BC, a manual of civil and structural engineering with extensive sections on materials and machines used in construction. One reason for their success is their accurate surveying techniques based on thedioptra, groma and chorobates.

Centuries later, in the 15th and 16th centuries and despite lacking beam theory and calculus, Leonardo da Vinci produced many engineering designs based on scientific observations and rigour, including a design for a bridge to span the Golden Horn. Though dismissed at the time, the design has since been judged to be both feasible and structurally valid

The foundations of modern structural engineering were laid in the 17th century by Galileo Galilei, Robert Hooke and Isaac Newtonwith the publication of three great scientific works.

In 1638 Galileo published *Dialogues Relating to Two New Sciences*, outlining the sciences of the strength of materials and the motion of objects (essentially defining gravity as a force giving rise to a constantacceleration).

It was the first establishment of a scientific approach to structural engineering, including the first attempts to develop a theory for beams. This is also regarded as the beginning of structural analysis, the mathematical representation and design of building structures.

This was followed in 1676 by Robert Hooke's first statement of Hooke's Law, providing a scientific understanding of elasticity of materials and their behaviour under load. Eleven years later, in 1687, Sir Isaac Newton published *Philosophiae Naturalis Principia Mathematica*, setting out his Laws of Motion, providing for the first time an understanding of the fundamental laws governing structures.

Also in the 17th century, Sir Isaac Newton and Gottfried Leibniz both independently developed the Fundamental theorem of calculus, providing one of the most important mathematical tools in engineering.

Further advances in the mathematics needed to allow structural engineers to apply the understanding of structures gained through the work of Galileo, Hooke and Newton during the 17th century came in the 18th century when Leonhard Euler pioneered much of the mathematics and many of the methods which allow structural engineers to model and analyse structures.

vSpecifically, he developed the Euler-Bernoulli beam equation withDaniel Bernoulli (1700–1782) circa 1750 - the fundamental theory underlying most structural engineering design.

Daniel Bernoulli, with Johann (Jean) Bernoulli (1667–1748), is also credited with formulating the theory of virtual work, providing a tool using equilibrium of forces and compatibility of geometry to solve structural problems. In 1717 Jean Bernoulli wrote to Pierre Varignon explaining the principle of virtual work, while in 1726 Daniel Bernoulli wrote of the "composition of forces". In 1757 Leonhard Euler went on to derive the Euler buckling formula, greatly advancing the ability of engineers to design compression elements.

MODERN DEVELOPMENTS IN STRUCTURAL ENGINEERING

Throughout the late 19th and early 20th centuries, materials science and structural analysis underwent development at a tremendous pace. Though elasticity was understood in theory well before the 19th century, it was not until 1821 that Claude-Louis Navier formulated the general theory of elasticity in a mathematically usable form.

In his *leçons* of 1826 he explored a great range of different structural theory, and was the first to highlight that the role of a structural engineer is not to understand the final, failed state of a structure, but to prevent that failure in the first place. In 1826 he also established the elastic modulus as a property of materials independent of the second moment of area, allowing engineers for the first time to both understand structural behaviour and structural materials.

Fig. Bessemer converter, Kelham Island Museum, Sheffield, England (2002)

Towards the end of the 19th century, in 1873, Carlo Alberto Castigliano presented his dissertation "Intorno ai sistemi elastici", which contains his theorem for computing displacement as partial derivative of the strain energy.In 1824, Portland cement was patented by the engineer Joseph Aspdin as *"a superior cement resembling Portland Stone"*, British Patent no. 5022. Although different forms of cement already existed (Pozzolanic cement was used by the Romans as early as 100 B.C. and even earlier by the ancient Greek and Chinese civilizations) and were in common usage in Europe from the 1750s, the discovery made by Aspdin used commonly available, cheap materials, making concrete construction an economical possibility.

Developments in concrete continued with the construction in 1848 of a rowing boat built of ferrocement - the forerunner of modernreinforced concrete - by Joseph-Louis Lambot. He patented his system of mesh reinforcement and concrete in 1855, one year after W.B. Wilkinson also patented a similar system. This was followed in 1867 when a reinforced concrete planting tub was patented by Joseph Monier in Paris, using steel mesh reinforcement similar to that used by Lambot and Wilkinson. Monier took the idea forward, filing several patents for tubs, slabs and beams, leading eventually to the Monier system of reinforced structures, the first use of steel reinforcement bars located in areas of tension in the structure.

Fig. Belper North Mill

Steel construction was first made possible in the 1850s when Henry Bessemer developed the Bessemer process to produce steel. He gained patents for the process in 1855 and 1856 and successfully completed the conversion of cast iron into cast steel in 1858.Eventually mild steel would replace both wrought iron and cast iron as the preferred metal for construction.

Fig. The Forth Bridge

During the late 19th century, great advancements were made in the use of cast iron, gradually replacing wrought iron as a material of choice. Ditherington Flax Mill in Shrewsbury, designed by Charles Bage, was the first building in the world with an interior iron frame. It was built in 1797. In 1792 William Strutt had attempted to build a fireproof mill at Belper in Derby (Belper West Mill), using cast iron columns and timber beams within the depths of brick arches that formed the floors. The exposed beam soffits were protected against fire by plaster. This mill at Belper was the world's first attempt to construct fireproof buildings, and is the first example of fire engineering. This was later improved upon with the construction of Belper North Mill, a collaboration between Strutt and Bage, which by using a full cast iron frame represented the world's first "fire proofed" building.

Fig. The Lattice shell structure of theShukhov Tower in Moscow.

The Forth Bridge was built by Benjamin Baker, Sir John Fowler and William Arrol in 1889, using steel, after the original design for the bridge by Thomas Bouch was rejected following the collapse of his Tay Rail Bridge. The Forth Bridge was one of the first major uses of steel, and a landmark in bridge design. Also in 1889, the wrought-iron Eiffel Tower was built by Gustave Eiffel and Maurice Koechlin, demonstrating the potential of construction using iron, despite the fact that steel construction was already being used elsewhere.

During the late 19th century, Russian structural engineer Vladimir Shukhov developed analysis methods for tensile structures, thin-shell structures, lattice shell structures and new structural geometries such as hyperboloid structures. Pipeline transport was pioneered byVladimir Shukhov and the Branobel company in the late 19th century.

Again taking reinforced concrete design forwards, from 1892 onwards François Hennebique's firm used his patented reinforced concrete system to build thousands of structures throughout Europe. Thaddeus Hyatt in the US and Wayss and Freitag in Germany also patented systems. The firm *AG für Monierbauten* constructed 200 reinforced concrete bridges in Germany between

1890 and 1897 The great pioneering uses of reinforced concrete however came during the first third of the 20th century, with Robert Maillart and others furthering of the understanding of its behaviour. Maillart noticed that many concrete bridge structures were significantly cracked, and as a result left the cracked areas out of his next bridge design - correctly believing that if the concrete was cracked, it was not contributing to the strength.

This resulted in the revolutionary Salginatobel Bridge design. Wilhelm Ritter formulated the truss theory for the shear design of reinforced concrete beams in 1899, and Emil Mörsch improved this in 1902. He went on to demonstrate that treating concrete in compression as a linear-elastic material was a conservative approximation of its behaviour. Concrete design and analysis has been progressing ever since, with the development of analysis methods such as yield line theory, based on plastic analysis of concrete (as opposed to linear-elastic), and many different variations on the model for stress distributions in concrete in compression

Prestressed concrete, pioneered by Eugène Freyssinet with a patent in 1928, gave a novel approach in overcoming the weakness of concrete structures in tension. Freyssinet constructed an experimental prestressed arch in 1908 and later used the technology in a limited form in the Plougastel Bridge in France in 1930. He went on to build six prestressed concrete bridges across the Marne River, firmly establishing the technology.

Structural engineering theory was again advanced in 1930 when Professor Hardy Cross developed his Moment distribution method, allowing the real stresses of many complex structures to be approximated quickly and accurately.

In the mid 20th century John Fleetwood Baker went on to develop the plasticity theory of structures, providing a powerful tool for the safe design of steel structures.

High-rise construction, though possible from the late 19th century onwards, was greatly advanced during the second half of the 20th century. Fazlur Khan designed structural systems that remain fundamental to many modern high rise constructions and which he employed in his structural designs for the John Hancock Center in 1969 and Sears Tower in 1973. Khan's central innovation in skyscraper design and construction was the idea of the "tube" and "bundled tube" structural systems for tall buildings

He defined the framed tube structure as "a three dimensional space structure composed of three, four, or possibly more frames, braced frames, or shear walls, joined at or near their edges to form a vertical tube-like structural system capable of resisting lateral forces in any direction by cantilevering from the foundation." Closely spaced interconnected exterior columns form the tube. Horizontal loads, for example wind, are supported by the structure as a whole. About half the exterior surface is available for windows. Framed tubes allow fewer interior columns, and so create more usable floor space. Where larger openings like garage doors are required, the tube frame must be interrupted, with transfer girders used to maintain structural integrity. The first building to apply the tube-frame construction was in the DeWitt-Chestnut Apartment Buildingwhich Khan designed in Chicago. This laid the foundations for the tube structures used in most later skyscraper constructions, including theconstruction of the World Trade Center.

Another innovation that Fazlur Khan developed was the concept of X-bracing, which reduced the lateral load on the building by transferring the load into the exterior columns. This allowed for a reduced need for interior columns thus creating more floor space, and can be seen in the John Hancock Center. The first sky lobby was also designed by Khan for the John Hancock Center in 1969. Later buildings with sky lobbies include the World Trade Center, Petronas Twin Towers and Taipei 101.

In 1987 Jörg Schlaich and Kurt Schafer published the culmination of almost ten years of work on the strut and tie method for concrete analysis - a tool to design structures with discontinuities such as corners and joints, providing another powerful tool for the analysis of complex concrete geometries.

In the late 20th and early 21st centuries the development of powerful computers has allowed finite element analysis to become a significant tool for structural analysis and design. The development of finite element Programmes has led to the ability to accurately predict the stresses in complex structures, and allowed great advances in structural engineering design and architecture. In the 1960s and 70s computational analysis was used in a significant way for the first time on the design of the Sydney Opera House roof. Many modern structures could not be understood and designed without the use of computational analysis.

Developments in the understanding of materials and structural behaviour in the latter part of the 20th century have been significant, with detailed understanding being developed of topics such as fracture mechanics, earthquake engineering, composite materials, temperature effects on materials, dynamics and vibration control, fatigue, creep and others. The depth and breadth of knowledge now available in structural engineering, and the increasing range of different structures and the increasing complexity of those structures has led to increasing specialisation of structural engineers.

STRUCTURAL ENGINEER

Structural engineers analyze, design, plan, and research structural components and structural systems to achieve design goals and ensure the safety and comfort of users or occupants. Their work takes account mainly of safety, technical, economic and environmental concerns, but they may also consider aesthetic and social factors. Structural engineering is usually considered a specialty discipline within civil engineering, but it can also be studied in its own right. In the US, most practicing structural engineers are currently licensed as civil engineers, but the situation varies from state to state. In the UK, most structural engineers in the building industry are members of the Institution of Structural Engineers rather than the Institution of Civil Engineers.

Typical structures designed by a structural engineer include buildings, towers, stadia and bridges. Other structures such as oil rigs, space satellites, aircraft and ships may also be designed by a structural engineer. Most structural engineers are employed in the construction industry, however there are also structural engineers in the aerospace, automobile and shipbuilding industries. In the construction industry, they work closely with architects, civil engineers, mechanical engineers, electrical engineers, quantity surveyors, and construction managers.

Structural engineers ensure that buildings and bridges are built to be strong enough and stable enough to resist all appropriate structural loads (*e.g.*, gravity, wind, snow, rain, seismic (earthquake), earth pressure, temperature, and traffic) in order to prevent or reduce loss of life or injury. They also design structures to be stiff enough to not deflect orvibrate beyond acceptable limits. Human comfort is an issue that is regularly considered in the limits. Fatigue is also an important consideration for bridges and for aircraft design or for other structures which experience a large number of stress cycles over their lifetimes. Consideration is also given to durability of materials against possible deterioration which may impair performance over the design lifetime.

EDUCATION

The education of structural engineers is usually through a civil engineering bachelor's degree, and often a master's degree specializing in structural engineering. The fundamental core subjects for structural engineering are strength of materials or solid mechanics, Structural Analysis -Static and Dynamic, material science and numerical analysis.Reinforced concrete, composite structure, timber, masonry and structural steel designs are the general structural design courses that will be introduced in the next level of the education of structural engineering. The structural analysis courses which include structural mechanics, structural dynamics and structural failure analysis are designed to build up the fundamental analysis skills and theories for structural engineering students. At the senior year level or in graduate Programmes, prestressed concrete design, space framedesign for building and aircraft, bridge engineering, civil and aerospace structure rehabilitation and other advanced structural engineering specializations are usually introduced.

Recently in the United States, there have been discussions in the structural engineering community about the knowledge base of structural engineering graduates. Some have called for a master's degree to be the minimum standard for professional licensing as a civil engineer. There are separate structural engineering undergraduate degrees at theUniversity of California, San Diego and at the University of Architecture, Civil Engineering and Geodesy, Sofia, Bulgaria. Many students who later become structural engineers major in civil, mechanical, or aerospace engineering degree Programmes, with emphasis in structural engineering. Architectural engineering Programmes do offer structural emphases, and are often in combined academic departments with civil engineering.

LICENSING OR CHARTERED STATUS

In the United States, persons practicing structural engineering must be licensed in each state in which they practice. Licensure may usually be obtained by the same qualifications as for a Civil Engineer, but some states require licensure specifically for structural engineering, with experience specific and non-concurrent with experience claimed for another engineering profession. The qualifications for licensure typically include a specified minimum level of practicing experience, as well as the successful completion of a nationally administered exam, and possibly a state-specific exam. For instance, California requires that candidates pass a national exam, written by the National Council of Examiners for Engineering and Surveying (NCEES), as well as a state-specific exam which includes a seismic portion and a surveying portion. Most states do not have a separate structural engineering license. In California, Washington, Oregon, Nevada, Illinois and other states, there is an additional license or authority for Structural Engineering, obtained after the engineer has obtained a Civil Engineering license and practiced an additional amount of time with the Civil Engineering license.

The scope of what may be designed by a Structural Engineer but not by a Civil Engineer without the S.E. license is very limited.

The United Kingdom has one of the oldest professional institutions for structural engineers. Originally founded as the Concrete Institute in 1908, it was renamed the Institution of Structural Engineers (IStructE) in 1922. It now has 22,000 members with branches in 32 countries.

The IStructE is one of several UK professional bodies empowered to grant the title of Chartered Engineer; its members are granted the title of Chartered Structural Engineer.

The overall process to become chartered begins after graduation from a UK MEng degree, or a BEng with an MSc degree. To qualify as a chartered structural engineer, a graduate needs to go through four years of Initial Professional Development followed by a professional review interview. After passing the interview, the candidate sits an eight-hour professional review examination.

The election to chartered membership (MIStructE) depends on the examination result. The candidate can register at the Engineering Council UK as a Chartered Structural Engineer once he or she has been elected as a

Chartered Member. Legally it is not necessary to be a member of the IStructE when working on structures in the UK, however industry practice, insurance and liabilities dictate that an appropriately qualified engineer be responsible for such work.

In the USA, application for license exam is allowed 4 years after the candidate graduated from an ABET accredited University and passing the fundamentals of Engineering exam, 3 years after receiving a master's degree, or 2 years after receiving a Ph.D. degree.

CAREER AND REMUNERATION

A 2010 survey of professionals occupying jobs in the construction industry showed that structural engineers in the UK earn an average wage of £35,009. The salary of structural engineers varies from sector to sector within the construction and built environment industry worldwide, depending on the project. For example, structural engineers working in public sector projects earn on average £37,083 per annum compared to the £43,947 average earned by those in commercial projects. Certain regions also represent higher average salaries, with structural engineers in the Middle East in all sectors, and of every level of experience, earning £45,083, compared to UK and EU countries where the average is £35,164.

ARCHITECTURAL ENGINEERING

Architectural engineering, also known as building engineering, is the application of engineering principles and technology to buildingdesign and construction. Definitions of an architectural engineer may refer to:

- An engineer in the structural, mechanical, electrical, construction or other engineering fields of building design and construction.
- A licensed engineering professional in parts of the United States.
- In informal contexts, and formally in some places, a professional synonymous with or similar to an architect.

ENGINEERING FOR BUILDING

STRUCTURAL ENGINEERING

Structural engineering involves the analysis and design of physical objects (buildings, bridges, equipment supports, towers and walls). Those concentrating on buildings are responsible for the structural performance of a large part of the built environment and are, sometimes, informally referred to as "building engineers". Structural engineers require expertise in strength of materials and in the seismic design of structures covered by earthquake engineering. Architectural Engineers sometimes practice structural as one aspect of their designs; the structural discipline when practiced as a specialty works closely with architects and other engineering specialists.

MECHANICAL, ELECTRICAL, AND PLUMBING (MEP)

Fig. MEP room in a building

Mechanical engineering and electrical engineering engineers are specialists, commonly referred to as "MEP" (mechanical, electrical, and plumbing) when engaged in the building design fields. Also known as "building services engineering" in the United Kingdom, Canada, andAustralia. Mechanical engineers often design and oversee the heating, ventilation and air conditioning (HVAC), plumbing, and rain guttersystems.

Plumbing designers often include design specifications for simple active fire protection systems, but for more complicated projects,fire protection engineers are often separately retained. Electrical engineers are responsible for the building's power distribution,telecommunication, fire alarm, signalization, lightning protection and control systems, as well as lighting systems.

THE ARCHITECTURAL ENGINEER (PE) IN THE UNITED STATES

In many jurisdictions of the United States, the architectural engineer is a licensed engineering professional. Usually a graduate of an architectural engineering university Programme preparing students to perform whole-building design in competition with architect-engineer teams; or for practice in one of structural, mechanical or electrical fields of building design, but with an appreciation of integrated architectural requirements.

Formal architectural engineering education, following the engineering model of earlier disciplines, developed in the late 19th century, and became widespread in the United States by the mid-20th century. With the establishment of a specific "architectural engineering" NCEES Professional Engineering registration examination in the 1990s, and first offering in April 2003, architectural engineering became recognized as a distinct engineering discipline in the United States. Architectural engineers are not entitled to practice architecture unless they are also licensed as architects.

THE ARCHITECT AS ARCHITECTURAL ENGINEER

In some countries, the practice of architecture includes planning, designing and overseeing the building's construction, and architecture, as a profession providing architectural services, is referred to as "architectural engineering". In Japan, a "first-class architect" plays the dual role of architect and building engineer, although the services of a licensed "structural design first-class architect" are required for buildings over a certain scale.

In some languages, such as Korean and Arabic, "architect" is literally translated as "architectural engineer". In some countries, an "architectural engineer" (such as the*ingegnere edile* in Italy) is entitled to practice architecture and is often referred to as an architect. These individuals are often also structural engineers. In other countries, such as Germany, Austria, Iran, and most of the Arabic countries, architecture graduates receive an engineering degree (*Dipl.-Ing. – Diplom-Ingenieur*).

In Spain, an "architect" has a technical university education and legal powers to carry out building structure and facility projects.

In Brazil, architects and engineers used to share the same accreditation process (CONFEA – Federal Council of Engineering, Architecture and Agronomy).

Now the Brazilian architects and urbanists have their own accreditation process (CAU – Architecture and Urbanism Council). Besides traditional architecture design training, Brazilian architecture courses also offer complementary training in engineering disciplines such as structural, electrical, hydraulic and mechanical engineering.

After graduation, architects can be fully responsible for design and construction in these areas (except in electric wiring, where the architect autonomy is limited to systems up to 30kVA), applied to buildings, urban environment, built cultural heritage, landscape planning, interiorscape planning and regional planning.

In Greece licensed architectural engineers are graduates from architecture faculties that belong to the Polytechnic University, obtaining an "Engineering Diploma".

They graduate after 5 years of studies and are fully entitled architects once they become members of the Technical Chamber of Greece (TEE – TεχVIKÓ EΠΙμεληTήρΙΟ Eλλάδς) .The Engineering Diploma equals a Master's Degree in ECTS units (300) according to the Bologna Accords.

EDUCATION

The architectural, structural, mechanical and electrical engineering branches each have well established educational requirements that are usually fulfilled by completion of a university Programme.

Fig. An air handling unit is used for the heating and cooling of air in a central location (click on image for legend). Bringing together knowledge of acoustic engineering and HVAC is one example of the multi-disciplined nature of architectural engineering

ARCHITECTURAL ENGINEERING AS A SINGLE INTEGRATED FIELD OF STUDY

What differentiates architectural engineering as a separate and single, integrated field of study, compared to other engineering disciplines, is its multi-disciplinary engineering approach. Through training in and appreciation of architecture, the field seeks integration of building systems within its overall building design. Architectural engineering includes the design of building systems including heating, ventilation and air conditioning (HVAC), plumbing, fire protection, electrical, lighting, architectural acoustics, and structural systems. In some university Programmes, students are required to concentrate on one of the systems; in others, they can receive a generalist architectural or building engineering degree.

WIND ENGINEERING

Wind engineering analyzes effects of wind in the natural and the built environment and studies the possible damage, inconvenience or benefits which may result from wind. In the field of structural engineering it includes strong winds, which may cause discomfort, as well as extreme winds, such as in a tornado, hurricane or heavy storm, which may cause widespread destruction. In the fields of wind energy and air pollution it also includes low and moderate winds as these are relevant to electricity production resp. dispersion of contaminants.

Wind engineering draws upon meteorology, fluid dynamics, mechanics, geographic information systems and a number of specialist engineering disciplines includingaerodynamics, and structural dynamics. The tools used include atmospheric models, atmospheric boundary layer wind tunnels, open jet facilities and computational fluid dynamics models.

Wind engineering involves, among other topics:

- Wind impact on structures (buildings, bridges, towers).
- Wind comfort near buildings.
- Effects of wind on the ventilation system in a building.
- Wind climate for wind energy.
- Air pollution near buildings.

Wind engineering may be considered by structural engineers to be closely related to earthquake engineering and explosion protection.

HISTORY

Wind Engineering as a separate discipline can be traced to the UK in the 1960s, when informal meetings were held at the National Physical Laboratory, the Building Research Establishment and elsewhere.

WIND LOADS ON BUILDINGS

The design of buildings must account for wind loads, and these are affected by wind shear. For engineering purposes, a power law wind speed profile may be defined as follows:

$$v_z = v_g \cdot \left(\frac{z}{z_g}\right)^{\frac{1}{\alpha}}, 0 < z < z_g$$

where:

v_z = speed of the wind at height z

v_g = gradient wind at gradient height z_g

α = exponential coefficient

Typically, buildings are designed to resist a strong wind with a very long return period, such as 50 years or more. The design wind speed is determined from historical records using extreme value theory to predict future extreme wind speeds.

WIND COMFORT

The advent of high rise tower blocks led to concerns regarding the wind nuisance caused by these buildings to pedestrians in their vicinity.

A number of wind comfort and wind danger criteria were developed from 1971, based on different pedestrian activities such as:

- Sitting for a long period of time
- Sitting for a short period of time
- Strolling
- Walking fast

Other criteria classified a wind environment as completely unacceptable or dangerous. Building geometries consisting of one and two rectangular buildings have a number of well-known effects:

- Corner streams, also known as corner jets, around the corners of buildings
- Through-flow, also known as a passage jet, in any passage through a building or small gap between two buildings due to pressure short-circuiting
- Vortex shedding in the wake of buildings

For more complex geometries, pedestrian wind comfort studies are required. These can use an appropriately scaled model in a boundary layer wind tunnel, or more recently there has been increased use of Computational Fluid Dynamics (CFD) techniques. The pedestrian level wind speeds for a given exceedance probability are calculated to allow for regional wind speeds statistics.

The vertical wind profile used in these studies varies according to the terrain in the vicinity of the buildings (which is may differ by wind direction), and is often grouped in categories such as:

- Exposed open terrain with few or no obstructions and water surfaces at serviceability wind speeds.
- Water surfaces, open terrain, grassland with few, well-scattered obstructions having heights generally from 1.5 m to 10m.
- Terrain with numerous closely spaced obstructions 3 m to 5 m high, such as areas of suburban housing.
- Terrain with numerous large, high (10 m to 30 m high) and closely spaced obstructions, such as large city centres and well-developed industrial complexes.

WIND TURBINES

Wind turbines are affected by wind shear. Vertical wind-speed profiles result in different wind speeds at the blades nearest to the ground level compared to those at the top of blade travel, and this in turn affects the turbine operation. The wind gradient can create a large bending moment in the shaft of a two bladed turbine when the blades are vertical. The reduced wind gradient over water means shorter and less expensive wind turbine towers can be used in shallow seas.

For wind turbine engineering, wind speed variation with height is often approximated using a power law:

$$v_w(h) = v_{ref} \cdot \left(\frac{h}{h_{ref}}\right)^a$$

where:

$\upsilon_w(h)$ = velocity of the wind at height h [m/s]

υ_{ref} = velocity of the wind at some reference height h_{ref} [m]

α = Hellman exponent (aka power law exponent or shear exponent) ($\sim$= 1/7 in neutral flow, but can be >1)

SIGNIFICANCE

The knowledge of wind engineering is used to analyze and design all high rise buildings, cable suspension bridges and cable-stayed bridges, electricity transmission towers and telecommunication towers and all other types of towers and chimneys.

The wind load is the dominant load in the analysis of many tall buildings. So wind engineering is essential for the analysis and design of tall buildings. Again, wind load is a dominant load in the analysis and design of all long-span cable bridges.

EARTHQUAKE RESISTANT STRUCTURES

Fig. Earthquake-proof and massive pyramid El Castillo, Chichen Itza.

Earthquake-resistant structures are structures designed to withstand earthquakes.

While no structure can be entirely immune to damage from earthquakes, the goal of earthquake-resistant construction is to erect structures that fare better during seismic activity than their conventional counterparts.

According to building codes, earthquake resistant structures are intended to withstand the largest earthquake of a certain probability that is likely to occur at their location.

This means the loss of life should be minimized by preventing collapse of the buildings for rare earthquakes while the loss of functionality should be limited for more frequent ones.

To combat earthquake destruction, the only method available to ancient architects was to build their landmark structures to last, often by making them excessively stiff and strong, like the El Castillo pyramid at Chichen Itza.

Currently, there are several design philosophies in earthquake engineering, making use of experimental results, computer simulations and observations from past earthquakes to offer the required performance for the seismic threat at the site of interest.

These range from appropriately sizing the structure to be strong and ductile enough to survive the shaking with an acceptable damage, to equipping it withbase isolation or using structural vibration control technologies to minimize any forces and deformations.

Fig. Snapshot from shake-table video of testing base-isolated (right) and regular (left) building model

While the former is the method typically applied in most earthquake-resistant structures, important facilities, landmarks and cultural heritage buildings use the more advanced (and expensive) techniques of isolation or control to survive strong shaking with minimal damage. Examples of such applications are the Cathedral of Our Lady of the Angels and the Acropolis Museum.

TRENDS, PROJECTS

Some of the new trends and/or projects in the field of earthquake engineering structures are presented below.

Proofed earthquake building material

A German composite construction company developed a proofed earthquake-safe supported core material based on internal beam/frame constructions. The same principle allows construction of hurricane-safe houses.

Earthquake shelter

One Japanese construction company has developed a six-foot cubical shelter, presented as an alternative to earthquake-proofing an entire building.

Concurrent shake-table testing

Concurrent shake-table testing of two or more building models is a vivid, persuasive and effective way to validate earthquake engineering solutions experimentally.

Thus, two wooden houses built before adoption of the 1981 Japanese Building Code were moved to E-Defence for testing. The left house was reinforced to enhance its seismic resistance, while the other one was not. These two models were set on E-Defence platform and tested simultaneously.

Combined vibration control solution

Fig. Close-up of abutment of seismically retrofitted Municipal Services Building in Glendale, CA

Fig. Seismically retrofitted Municipal Services Building in Glendale, CA

Designed by architect Merrill W. Baird of Glendale, working in collaboration with A. C. MartinArchitects of Los Angeles, the Municipal Services Building at 633 East Broadway, Glendale was completed in 1966. Prominently sited at the corner of East Broadway and Glendale Avenue, this civic building serves as a heraldic element of Glendale's civic center.

In October 2004 Architectural Resources Group (ARG) was contracted by Nabih Youssef and Associates, Structural Engineers, to provide services regarding a historic resource assessment of the building due to a proposed seismic retrofit.

In 2008, the Municipal Services Building of the City of Glendale, California was seismically retrofitted using an innovative combined vibration control solution: the existing elevated building foundation of the building was put on high damping rubber bearings.

Steel plate shear walls system

Fig. Coupled steel plate shear walls,Seattle, Copyright Mehdi Kharrazi

Fig. The Ritz-Carlton/JW Marriott hotel building engaging the advanced steel plate shear walls system, LA

A steel plate shear wall (SPSW) consists of steel infill plates bounded by a column-beam system. When such infill plates occupy each level within a framed bay of a structure, they constitute a SPSW system. Whereas most earthquake resistant construction methods are adapted from older systems, SPSW was invented entirely to withstand seismic activity.

SPSW Behaviour is analogous to a vertical plate girder cantilevered from its base. Similar to plate girders, the SPSW system optimizes component performance by taking advantage of the post-buckling Behaviour of the steel infill panels.

The Ritz-Carlton/JW Marriott hotel building, a part of the LA Live development in Los Angeles, California, is the first building in Los Angeles that uses an advanced steel plate shear wall system to resist the lateral loads of strong earthquakes and winds.

Kashiwazaki-Kariwa Nuclear Power Plant is partially upgraded

The Kashiwazaki-Kariwa Nuclear Power Plant, the largest nuclear generating station in the world by net electrical power rating, happened to be near the epicenter of the strongest M_w 6.6 July 2007 Chûetsu offshore earthquake. This initiated an extended shutdown for structural inspection which indicated that a greater earthquake-proofing was needed before operation could be resumed.

On May 9, 2009, one unit (Unit 7) was restarted, after the seismic upgrades. The test run had to continue for 50 days. The plant had been completely shut down for almost 22 months following the earthquake.

Seismic Test of Seven-Story Building

A destructive earthquake struck a lone, wooden condominium in Japan. The experiment was webcast live on July 14, 2009 to yield insight on how to make wooden structures stronger and better able to withstand major earthquakes.

The Miki shake at the Hyogo Earthquake Engineering Research Center is the capstone experiment of the four-year NEESWood project, which receives its primary support from the U.S. National Science Foundation Network for Earthquake Engineering Simulation (NEES) Programme.

"NEESWood aims to develop a new seismic design philosophy that will provide the necessary mechanisms to safely increase the height of wood-frame structures in active seismic zones of the United States, as well as mitigate earthquake damage to low-rise wood-frame structures," said Rosowsky, Department of Civil Engineering at Texas AandM University. This philosophy is based on the application of seismic damping systems for wooden buildings. The systems, which can be installed inside the walls of most wooden buildings, include strong metal frame, bracing and dampers filled with viscous fluid.

4

The Designing and Construction of Civil Engineering

DESIGN AND CONSTRUCTION AS AN INCORPORATED ORGANIZATION

In the designing of facilities, it is significant to recognize the close relationship between conception and expression. These processes can best be viewed as an integrated organization. Broadly speaking, design comprises a process of creating the description of a new facility, generally represented by detailed plans and specifications; construction planning is a process from identifying activities and resources required to make the conception a physical reality. Hence, construction is the execution of a design envisioned by architects and engineers. In some design and construction, numerous operational tasks must be executed with a variety of precedence and other relationships among the different tasks.

Several characteristics are unique to the planning of constructed facilities and should be kept in mind even at the very early stage of the project life cycle. These include the following:

- Nearly every facility is custom designed and constructed, and often requires a long time to complete.
- Both the design and construction of a facility must satisfy the conditions peculiar to a specific site.
- Because each project is site specific, its execution is influenced by natural, social and other locational conditions such as weather, Labour supply, local building codes, etc.
- Since the service life of a facility is long, the anticipation of future requirements is inherently difficult.
- Because of technological complexity and market demands, changes of design plans during construction are not uncommon.

In an integrated system, the preparation for both design and construction can proceed almost simultaneously, examining various alternatives which are desirable from both viewpoints and thus eliminating the necessity of extensive revisions under the guise of value engineering. Furthermore, the review of designs with regard to their constructibility can be carried out as the project progresses from planning to design. For example, if the sequence of assembly of a structure and the critical loadings on the partially assembled structure during construction are carefully considered as a part of the overall structural design, the impacts of the design on construction falsework and on assembly details can be anticipated.

However, if the design professionals are expected to assume such responsibilities, they must be rewarded for sharing the risks as well as for undertaking these additional tasks. Similarly, when construction contractors are expected to take over the responsibilities of engineers, such as devising a very elaborate scheme to erect an unconventional structure, they too must be rewarded accordingly. As long as the owner does not assume the responsibility for resolving this risk-reward dilemma, the concept of a truly integrated system for the design and construction cannot be realized.

It is motivating to note that European owners are generally more open to new technologies and to share risks with designers and contractors. In particular, they are more willing to accept responsibilities for the unforeseen subsurface conditions in Geotechnical engineering. Consequently, the designers and contractors are also more willing to introduce new techniques in order to reduce the time and cost of construction. In European practice, owners typically present contracts with a conceptual design, and contractors prepare detailed designs, which are checked by the owner's engineers. Those detailed designs may be alternate designs, and specialty contractors may also prepare detailed alternate designs.

ACCOUNTABILITY FOR SHOP DRAWINGS

The willingness to presuppose responsibilities does not come easily from any party in the current litigious climate of the construction industry in the United States. On the other hand, if owner, architect, engineer, contractor and other groups that represent parts of the industry do not jointly fix the responsibilities of various tasks to appropriate parties, the standards of practice will eventually be set by court decisions. In an attempt to provide a guide to the entire spectrum of participants in a construction project.

Shop drawings represent the assembly details for erecting a structure which should reflect the intent and rationale of the original structural design. They are prepared by the construction contractor and reviewed by the design professional. However, since the responsibility for preparing shop drawings was traditionally assigned to construction contractors, design professionals took the view that the review process was advisory and assumed no responsibility for their accuracy.

This justification was ruled unacceptable by a court in connection with the walkway failure at the Hyatt Hotel in Kansas City in 1985. In preparing the ASCE Manual of Professional Practice for Quality in the Constructed Project, the responsibilities for preparation of shop drawings proved to be the most difficult to develop.

The reason for this situation is not difficult to fathom since the responsibilities of the task are diffused, and all parties must agree to the new responsibilities assigned to each in the recommended risk-reward relations.

Traditionally, the owner is not involved in the preparation and review of shop drawings, and perhaps is even unaware of any potential problems. In the recommended practice, the owner is required to take responsibility for providing adequate time and funding, including approval of scheduling, in order to allow the design professionals and construction contractors to perform satisfactorily.

MODEL METRO PROJECT IN MILAN, ITALY

Under Italian law, unforeseen subsurface conditions are the owner's responsibility, not the contractor's. This is a striking difference from U.S. construction practice where changed conditions clauses and claims and the adequacy of Prebid site investigations are points of contention. In effect, the Italian law means that the owner assumes those risks. But under the same law, a contractor may elect to assume the risks in order to lower the bid price and thereby beat the competition.

According to the Technical Director of Rodio, the Milan-based contractor which is heavily involved in the grouting job for tunneling in the Model Metro project in Milan, Italy, there are two typical contractual arrangements for specialized subcontractor firms such as theirs. One is to work on a unit price basis with no responsibility for the design. The other is what he calls the "nominated subcontractor" or turnkey method: prequalified subcontractors offer their own designs and guarantee the price, quality, quantities, and, if they wish, the risks of unforeseen conditions.

At the beginning of the Milan metro project, the radio contract ratio was 50/50 unit price and turnkey. The firm convinced the metro owners that they would save money with the turnkey approach, and the ratio became 80per cent turnkey.

What's more, in the work packages where radio worked with other grouting specialists, those subcontractors paid radio a fee to assume all risks for unforeseen conditions.

Under these circumstances, it was critical that the firm should know the subsurface conditions as precisely as possible, which was a major reason why the firm developed a computerized electronic sensing Programme to predict stratigraphy and thus control grout mixes, pressures and, most important, quantities.

MODERNIZATION AND TECHNOLOGICAL FEASIBILITY

The planning for a construction project begins with the generation of concepts for a facility which will meet market demands and owner needs. Innovative concepts in design are highly valued not for their own sake but for their contributions to reducing costs and to the improvement of aesthetics, comfort or convenience as embodied in a well-designed facility. However, the constructor as well as the design professionals must have an appreciation and full understanding of the technological complexities often associated with innovative designs in order to provide a safe and sound facility. Since these concepts are often preliminary or tentative, screening studies are carried out to determine the overall technological viability and economic attractiveness without pursuing these concepts in great detail. Because of the ambiguity of the objectives and the uncertainty of external events, screening studies call for uninhibited innovation in creating new concepts and judicious Judgement in selecting the appropriate ones for further consideration.

One of the most important aspects of design innovation is the necessity of communication in the design/construction partnership. In the case of bridge design, it can be illustrated by the following quotation from Lin and Gerwick concerning bridge construction: The great pioneering steel bridges in the United States were built by an open or a covert alliance between designers and constructors. The turnkey approach of the designer - constructor has developed and built our chemical plants, refineries, steel plants, and nuclear power plants. It is time to ask, seriously, whether we may not have adopted a restrictive approach by divorcing engineering and construction in the field of bridge construction.If a contractor-engineer, by some stroke of genius, were to present to design engineers today a wonderful new scheme for long span pre-stressed concrete bridges that made them far cheaper, he would have to make these ideas available to all other constructors, even limiting or watering them down so as to "get a group of truly competitive bidders." The engineer would have to make sure that he found other contractors to bid against the ingenious innovator.

If an engineer should, by a similar stroke of genius, hit on such a unique and brilliant scheme, he would have to worry, wondering if the low bidder would be one who had any concept of what he was trying to accomplish or was in any way qualified for high class technical work.

Innovative design concepts must be tested for technological feasibility. Three levels of technology are of special concern: technological requirements for operation or production, design resources and construction technology. The first refers to the new technologies that may be introduced into a facility which is used for a certain type of production such as chemical processing or nuclear power generation. The second refers to the design capabilities that are available to the designers, such as new computational methods or new materials. The third refers to new technologies which can be adopted to construct the facility, such as new equipment or new construction methods.

A new facility may involve complex new technology for operation in hostile environments such as severe climate or restricted accessibility. Large projects with unprecedented demands for resources such as Labour supply, material and infrastructure may also call for careful technological feasibility studies. Major elements in a feasibility study on production technology should include, but are not limited to, the following:

- Project type as characterized by the technology required, such as synthetic fuels, petrochemicals, nuclear power plants, etc.
- Project size in dollars, design engineer's hours, construction Labour hours, etc.
- Design, including sources of any special technology which require licensing agreements.
- Project location which may pose problems in environmental protection, Labour productivity and special risks.

An example of innovative design for operation and production is the use of entropy concepts for the design of integrated chemical processes. Simple calculations can be used to indicate the minimum energy requirements and the least number of heat exchange units to achieve desired objectives. The result is a new incentive and the criterion for designers to achieve more effective designs. Numerous applications of the new methodology have shown its efficacy in reducing both energy costs and construction expenditures. This is a case in which innovative design is not a matter of trading-off operating and capital costs, but better designs can simultaneously achieve improvements in both objectives.

The choice of construction technology and method involves both *strategic* and *tactical* decisions about appropriate technologies and the best sequencing of operations. For example, the extent to which prefabricated facility components will be used represents a *strategic* construction decision. In turn, prefabrication of components might be accomplished off-site in existing manufacturing facilities or a temporary, on-site fabrication plant might be used. Another example of a strategic decision is whether to install mechanical equipment in place early in the construction process or at an intermediate stage. The strategic decisions of this sort should be integrated with the process of facility design in many cases. At the tactical level, detailed decisions about how to accomplish particular tasks are required, and such decisions can often be made in the field.

Construction planning should be a major concern in the development of facility designs, in the preparation of cost estimates, and in forming bids by contractors. Unfortunately, planning for the construction of a facility is often treated as an afterthought by design professionals. This contrasts with manufacturing practices in which the *assembly* of devices is a major concern in design. Design to insure ease of assembly or construction should be a major concern of engineers and architects. As the Business Roundtable noted, "All too often chances to cut schedule time and costs are lost because construction operates as a production process separated by a chasm from financial planning, scheduling, and engineering or architectural design. Too many engineers, separated from field experience, are not up to date about how to build what they design, or how to design so structures and equipment can be erected most efficiently."

CONSTRUCTION ENGINEERING

Fig. Berlin Brandenburg Airport, an example for poor construction planning and execution.

Construction engineering is a professional discipline that deals with the designing, planning, construction, andmanagement of infrastructures such as highways, bridges, airports, railroads, buildings, dams, and utilities. These Engineers are unique such that they are a cross between civil engineers and construction managers. Construction engineers learn the designing aspect much like civil engineers and construction site management functions much like construction managers.

The primary difference between a construction engineer and a construction manager is that the construction engineer has the ability to sit for the Professional Engineer license (PE) whereas a construction manager cannot. At the educational level, construction managers are not as focused on design work as they are on construction procedures, methods, and people management. Their primary concern is to deliver a project on time, within budget, and of the desired quality.

The difference between a construction engineer and civil engineer is only at the educational level as both disciplines are able to sit for the PE exam giving them the same title of engineer. Civil engineering students concentrate more on the design work, gearing them Towards a career as a design professional. This essentially requires them to take a multitude of design courses. Construction engineering students take design courses as well as construction management courses. This allows them to understand both the design functions as well as the building requirements needed to design and build today's infrastructures.

WORK ACTIVITIES

Depending on which career the construction engineer has chosen to follow, an entry-level design engineer normally provides support to project managers and assist with creating conceptual designs, scopes, and cost estimates for the planning and construction of approved projects. It should be noted that a career in design work does require a professional engineer license (PE). Individuals who pursue this career path are strongly advised to sit for the Engineer In Training exam (EIT) while in college as it takes five years (4 years in USA) post graduate to obtain the PE license.

Entry-level construction manager positions are typically called project engineers or assistant project engineers. They are responsible for preparing purchasing requisitions, processing change orders, preparing monthly budgeting reports, and handling meeting minutes. The construction management position does not necessarily require a PE license; however possessing one does make the individual more marketable, as the PE license allows the individual to sign off on temporary structure designs.

ABILITIES

Construction engineers are problem solvers, they help create infrastructure that best meets the unique demands of its environment. They must be able to understand infrastructure life cycles and have the perspective to solve technical challenges with clarity and imagination. Therefore individuals should have a strong understanding of maths and science, but many other skills are required, including critical and analytical thinking, time management, people management and good communication skills.

EDUCATIONAL REQUIREMENTS

Individuals looking to obtain a construction engineering degree must first ensure that the Programme is accredited by EAC or Technology Accreditation Commission (TAC) of theAccreditation Board for Engineering and Technology (ABET). ABET accreditation is assurance that a college or university Programme meets the quality standards established by the profession for which it prepares its students. In the US there are currently twenty-five Programmes that exist in the entire country so careful college consideration is advised.

A typical construction engineering curriculum is a mixture of engineering mechanics, engineering design, construction management and general science and mathematics. This usually leads to a Bachelor of Science degree. The B.S. degree along with some design or construction experience is sufficient for most entry level positions. Graduate schools may be an option for those who want to go further in depth of the construction and engineering subjects taught at the undergraduate level. In most cases construction engineering graduates look to either civil engineering, engineering management, or business administration as a possible graduate degree.

JOB PROSPECTS

Job prospects for construction engineers generally have a strong cyclical variation. For example, starting in 2008 - continuing until at least 2011 - job prospects have been poor due to the collapse of housing bubbles in many parts of the world. This sharply reduced demand for construction, forced construction professionals towards infrastructure construction and therefore increased the competition faced by established and new construction engineers. This increased competition, and a core reduction in quantity demand is in parallel with a possible shift in the demand for construction engineers due to the automation of many engineering tasks, overall resulting in reduced prospects for construction engineers. In early 2010 the United States construction industry had a 27per cent unemployment rate, this is nearly three times higher than the 9.7per cent national average unemployment rate. The construction unemployment rate (including tradesmen) is comparable to the United States 1933 unemployment rate - the lowest point of the Great Depression - of 25per cent.

REMUNERATION

The average salary for a civil engineer in the UK depends on the sector, and more specifically the level of experience of the individual. A 2010 survey of the remuneration and benefits of those occupying jobs in construction and the built environment industry showed that the average salary of a civil engineer in the UK is £29,582. In the United States, as of May 2013, the average was $85,640. The average salary varies depending on experience, for example the average annual salary for a civil engineer with between 3 and 6 years experience is £23,813. For those with between 14 and 20 years experience the average is £38,214.

INFORMATION TECHNOLOGY IN CONSTRUCTION ENGINEERING

The analysis of entropy engineering applications in construction comprises a young field of explore, still struggling to define its place within the larger category of academic disciplines. Being a young branch of science, information technology in construction (for which the abbreviation ITC will comprise used in the following text) lacks a solid methodological foundation. This is in contrast to some older engineering disciplines which are based on basic sciences such as physics and mathematics, and where testing can be carried out in a systematic fashion in laboratory conditions. The only paradigm that most researchers in the ITC domain currently share seem to be "object-orientation", a term which can be given many shades of meaning, depending on the context. Other than that there is a multitude of different research directions ranging from computer programming to management strategies. Practitioners and researchers alike are offered a wide range of IT techniques and management philosophies, many of which claim to be the ideal solution to the industry's problems. Current and recent buzzwords include knowledge-based systems, product data technology, the Internet, EDI, as well as concurrent engineering, lean construction, business process reengineering, total quality management, supply-chain management and just-in-time production.

Some generally accepted guidelines for how researchers can prove their "hypotheses" are needed. Some of the standard scientific techniques which all doctoral students are supposed to learn as a part of the training (*i.e.* replicability of experiments based on the information given in a thesis or paper, the statistical basis for proofs of the validity of models), are rarely rigorously applied in much of the reported ITC research.

It is difficult to give a very precise definition of the domain of ITC and to draw crystal clear boundaries between ITC and nearly related research domains. Often the discussion of IT technologies of interest to construction is centered on the most recent tools that general developments in commercial IT or in Computer science research have to offer (a "technology push" viewpoint). Good examples are object-orientation, world wide web, expert systems. A contrasting viewpoint would be to study the information management process in construction in a comprehensive way and to identify potential application areas for IT tools (a "problem driven" approach).

OPTIONS FOR DEFINING THE DOMAIN OF ITC

In principle, there are thus at smallest amount two options for defining the domain in a systematic way; a bottom-up bibliographical analysis of what researchers are actually doing or a top-down analysis based on some model of information management in construction. According to the first option it would be possible to provide a "map" of ITC research through a bibliographical analysis of the topics covered in the papers to be found in the leading ITC journals and conference proceedings, or by studying the contents of some databases of research projects in the domain. There are probably a few hundred researchers worldwide who are active within this field. Implicitly they have classified themselves as belonging in this field by submitting their articles to the half-dozen journals which explicitly deal with ITC topics (*i.e.* ASCE Journal of Computing in Civil Engineering, Automation in Construction, International

Journal of Construction Information Technology, Electronic Journal of Information Technology in Construction, International Journal of Computer-Integrated Design and Construction, Microcomputers in Civil Engineering) or by attending the limited number of annual conferences in the domain. Such a bibliographical analysis could be carried out with a moderate effort. Probably a few topics (*i.e.* expert systems, product modelling and recently web technology) would stand out in such an analysis. The drawback is that areas of application of IT in construction, which in practice are quite important, would be poorly represented since researchers have been relatively uninterested in them (for example Document management, EDI).

In this paper the second option, using a highly abstracted model of information management in construction as the basis for a definition, is used. In order to arrive at such as a model we first need a clear understanding of what we mean by "information technology" and "construction" and of the relationship between these two.

EXPLANATION OF "CONSTRUCTION AND "INFORMATION TECHNOLOGY"

It seems suitable to start with construction since this is the fundamental activity to which IT techniques are applied. The purpose of construction activities is to produce artifacts such as buildings, process plants, roads and bridges. Civil engineering artifacts are, in contrast to most other manufactured products, located in particular places and need to be constructed on-site rather than in factories. They are also usually one-of-a-kind products. The duration of a construction project is usually long. A comprehensive definition of the construction process should clearly include the whole life-cycle of civil engineering artefacts, including both design, construction, operation and maintenance. In particular, it is important to stress the inclusion of operation and maintenance since an important part of the information used during these stages originates during design and construction. It is also important to include the manufacturing of the building materials needed as well as public planning and inspection activities, activities which often are overlooked in process models of construction.

Information technology (IT) can be distinct as the use of electronic machines and Programmes for the processing, storage, transfer and presentation of information. In earlier days, when the emphasis was on processing the term electronic data processing, EDP, was common. Nowadays the use of information technology is no longer confined to huge number-crunching machines housed in air-conditioned computer halls but permeates all aspects of everyday life. Communications technology is today an important part of IT. Not only computers and their software, but also devices such as the telephone, the photocopying machine and the telefax should thus be included in our definition of information technology. Many of the functions of these devices are in fact increasingly integrated. With the latest generation of laptop computers it is already possible to send and receive faxes and e-mails. Recently, mobile phones which incorporate small microcomputers have started to appear on the market.

A SIMPLIFIED REPRESENTATION OF THE CONSTRUCTION PROCESS

The construction process: two interacting subprocesses

The modelling technique is used in some of the figures. IDEF0 is not the ideal modelling tool for this purpose, but despite some deficiencies and limitations it is easy to understand and there are good computer-based modelling tools available. An IDEF0 model consists of a number of boxes representing activities. Each activity takes some *inputs* (such as information, raw materials, etc.) and transforms these into *outputs* (information, buildings and products). An activity is performed by actors with the help of machines, computer software, etc. The latter is called mechanisms and are shown as arrows underneath the activity box. An activity is on a more abstract level controlled by instructions or more general knowledge (*controls).*

In a highly abstract way the construction process can be divided into two highly integrated sub-processes which interact with each other at many different levels. This subdivision is based on the nature of the objects that these sub-processes deal with. The information sub-process activities always result in information whereas the material sub-process activities produce services of physical objects.

In the material process raw materials and prefabricated components are created, modified, moved and installed and finally become embedded parts of the finished artefact. If we were to film a construction site during its whole duration, and to show it in extreme "fast-motion", what we would observe is almost exclusively the material process. But the material process cannot function on its own.

In contrast to the physical or chemical processes occurring in nature, the creation of any man-made artefact requires an information process which initiates and controls the necessary material activities. The immediate results of the information process are presented as drawings, specifications, schedules, procurement orders, etc. which control all material activities either by specifying the resulting artefact (design information) or the activities that need to be carried out in order for the artefact to be constructed (management information). Both types of activities utilise resources which are consumed in the process (materials, energy, labour, wear and depreciation of machinery). The cost of an activity is the direct result of the consumption of resources. A special type of resource or input is information, which as such is not consumed in the process of using it, but nevertheless has a price.The information and material sub-processes are integrated by information flows in two directions. Firstly, the information process produces information which indirectly or directly controls the material activities taking place. Secondly, the information processing activities constantly need feedback information about what is actually happening in the material process, in order to check compliance with the designs or monitor the progress of the work against the schedules. In a longer time perspective the information process also needs feedback on the performance of buildings during the maintenance stages.

The interface between control in sequence and actions in the physical world is consequently of interest. On one hand, information needs to be transformed into actions carried out by persons or by persons aided by tools and machines. The extreme case is the use of robotics, where information on higher levels needs to be transformed into very detailed instruction for how the robot moves its arms. Going in the other direction, physical impulses such as temperatures, pressures, light, etc. need to be transformed into information using measurement equipment. The simplest transformation is done by the human eye and brain by visual observation. This mechanism can today in many cases be substituted by IT-enabled techniques such as bar code readers and automatic pattern recognition.

Levels of concept in the model process

This clear split into information and material process performance can be observed on several levels of abstraction. On the highest level in our schematic model we can envisage the whole construction process, say from inception to the delivery and use of the finished artefact, which consists of a higher part aggregating the set of all information processing activities and a lower part containing all material handling activities.

If we look at the process slightly more detail we will notice that the information process includes several consecutive and parallel activities even before any material activities start. Stages such as briefing, schematic design, tendering, etc. are basically only information processing activities and it is only in the later stages of the construction process that there is a close correspondence between information and material activities, in the sense that the information process results in detailed instructions which control material activities. The end results from an earlier information processing activity are used as input information to the next activity (for instance, the client's brief as input to the schematic design stage, and the architect's designs as the basis for structural and building services design). In addition to product and process definition activities, analysis activities which aim at predicting aspects of the material process during construction or operation of the facility (*i.e.* cost estimation, energy simulation, FEM-analysis) are typical in these early stages.

In later life-cycle stages, we discover that the information activities start to result in information which more directly is used to control the material process, such as the procurement of materials and the construction activities on the site.

At some point of detailing in our hierarchical model (*i.e.* weekly or daily planning of on-site activities) a stage is reached where the formalised, often company-specific documentation routines end and where oral communication starts. Much of the detailed information processing is left to the individual workers actually carrying out the tasks. This does not mean that the information process ends, the basic abstraction is still valid, but that at this level of detailing formal documents are no longer produced. A human bricklayer for instance does not require as detailed instructions as a robot doing the same job would.

A significant trend of the last hundred years or so is, nevertheless, that more and more of the information needed is explicitly formulated in project documents, master specifications, etc. In earlier centuries much of the information was conveyed orally and there was heavy reliance on craftsmanship. Part of this increased degree of formalisation could also be explained by the more and complicated building services systems which need to be documented, but for the most part the underlying reason seems to be the increased division of labour in the construction industry.

The subdivision into material and information activities has been discussed quite at length above. One reason for this is that this way of describing the process differs somewhat from the mainstream of a construction process modelling efforts. In many of the descriptions found in the literature, the modelling tends more to follow existing organizational borders and current documentation practice. In such models the design phases are clearly distinguished, but it is difficult to separate out the information handling activities involved in site construction and the procurement of materials.

TRENDS IN THE EXERCISE OF MACHINES IN CONSTRUCTION

In a chronological perspective, the construction industry shows an increasing trend to use machines to automate both the material and the information processing activities. Since the industrial revolution started in the 18th century, machines have been used to automate or to aid man in performing material handling tasks. Tremendous increases in productivity have been achieved in particular in the large scale movement of materials typical of infrastructure projects.

Since the latter half of the 20th century machines have increasingly been used also to aid in information processing tasks. Early uses were in particular computer applications for engineering analysis. Since the 1980's IT use, in the form of CAD and word processing software, copying machines, faxes, mobile phones, computer networks, etc. has increased enormously and now affects all aspects of the information process.There are two different ways of using machines for automating activities (or for aiding humans in performing them). The first one is to take the manual process as it stands and use machines or computers to enhance it (straightforward automation). The second option is to redesign the process, taking into account the possibilities offered by machines (re-engineering).

A good instance of straightforward automation is the use of a CAD-drafting Programme instead of a drawing board for the production of construction drawings. A CAD-system certainly offers some productivity gains compared to the traditional process, in particular for managing change or if repetitive drawing elements are used. The end result is, nevertheless, almost exactly the same as in manual drafting and often the resulting drawings are copied and mailed just like before.

An instance of re-engineering is provided by the way the Swedish facilities management company ABB Fastigheter uses IT to enable service personnel to handle complaints and malfunctions related to the thermal comfort and energy usage in their large building stock. The service personnel are contacted through their mobile phones. Instead of rather expensive travel to correct situations, in particular during weekends and in the sparsely populated northern parts of Sweden, they access the computerized control system of the building in question over a modem from their laptop computers and make any necessary adjustments remotely. This leads to substantial reductions in the amount of personnel and travel needs and also makes the work itself more comfortable.

IT has, in the original stages of its introduction in the construction industry, mostly being used for straightforward automation. Only after a number of years have enterprises learnt about the opportunities offered by IT and started to use IT in more innovative ways. The recent developments in networking and communications technology and the miniaturization of the hardware have also started to offer increasing possibilities for re-engineering.

RISK-BASED COST CONTROL FOR CONSTRUCTION

The primary focus of this paper is to present a methodology to assist project managers and cost engineers to control cost when constructing a complex structure. Cost control during the construction process is vital to ensure the success of a project. The construction of a complex structure is a major undertaking and typically has people specifically responsible for cost control. Most complex projects share the common theme that they are fraught with risks and uncertainty that can cause cost escalations. For example, Laufer and Howell found that about 80per cent of all projects begin the construction process with a high level of uncertainty. Construction project manager and cost engineers have a challenging task to build a complex project that is on budget. Construction and cost engineering professionals have long recognised the need for improvements in the area of cost control. Managing costs includes estimating, scheduling, accumulating and analysing cost data, and finally implementing measures to correct a cost

problem. There are several cost control techniques that are used by the construction industry to varying degrees. These are: exception reporting, trend analysis, earned value, range estimating, and forecast unit cost. Except for perhaps earned value, these cost control techniques tend to focus on variances in line items once the cost overrun has been discovered. What is needed is a cost control methodology that proactively seeks out potential cost issues and provides project managers with as much warning as possible before their occurrence.

RISK ANALYSIS

Risk analysis is identified as a major subset to project management. However, its application appears limited within the field of construction project management. It is limited in its widespread use, sometimes it is only applied to developing schedules, or cost estimates. The reasons for this limited use are difficult to identify and quantify.

McKim proposed that risk analysis is viewed by construction cost engineers as an intimidating subject. Blair states that the data required to apply risk analysis is often expressed in linguistic terms and is difficult to apply in a classic quantitative risk analysis.

An application of fuzzy logic and set theory can be used as a solution to the later reason, and. Al-Bahar and Crandall also point out that construction professionals tend to use rules of thumb, rely on intuition, or experience when dealing with risk.

What is needed is an application of risk analysis to help project managers control cost that is relatively simple to apply, can be used throughout the life cycle of a construction project, accounts for the tendency of construction professionals to apply risk in linguistic terms, and apply their experience.

Risk Defined

The literature abounds with definitions of risk. Risk can be a somewhat ambiguous term unless its definition and convention are clearly stated. The concept of risk is used to assess and evaluate uncertainties associated with an event. Risk can be measured as a pair of the probability of occurrence (likelihood) of an event and the consequences (outcomes) associated with the event's occurrence. This pairing is not a mathematical operation, a scalar or vector quantity, but a matching of an event's likelihood of occurrence with the expected outcome. The generally accepted expression for risk is shown in Equation 1.

$$Risk \equiv [(P_1, C_1),(P_2, C_2),...,(P_x, C_x)]$$

In this equation P_x is the occurrence probability of event x, and C_x is the occurrence consequences or outcomes of event x. The consequences of a risk event in the project management field will most likely be in terms of dollars. Risk is sometimes thought as the potential for harm. In a project management context and as defined in this paper risk is thought of as being concerned with opportunities or potential gain as well as negative consequences. This broader risk definition is known as risk engineering.

There is a consensus within the technical community that a comprehensive risk analysis consists of risk assessment, risk management, and risk communication. Risk assessment is the process of identifying and evaluating areas of risk. Risk management is the act or practice of dealing with or controlling this risk. Risks can be managed based on the information provided from the risk assessment to determine risk acceptability and using decision analysis to make risk informed decisions for control and mitigation purposes. Risk communication is the process of documenting and exchanging information about the results of risk studies to various interested parties.

Risk Assessment

Risk assessment is a technical and scientific process by which the risk of given situations for a system are modeled and quantified. Risk assessment provides qualitative and quantitative data to decision-makers for later use in risk management. Risk assessment for construction projects can be performed by comparing the resource requirements needed to build the projects to the existing industrial capacity and by performing simulations of the construction processes. These techniques highlight the critical cost areas and opportunities for savings. When data does not exist or is unavailable, a construction risk assessment can be made in qualitative terms. Where data exists or can be obtained, the risk assessment is quantitative. The risk assessment then considers deviations from these construction scenarios that can lead to undesirable or positive consequences. The consequences can be described in terms of adverse or positive impacts to a project's cost, schedule, safety, or technical performance.

Qualitative risk analysis uses expert opinion to evaluate the probability and consequence of an event's occurrence. Safety review/ audit, checklist, what-if, Hazard and Operability study (HAZOP), Preliminary Hazard Analysis (PrHA), risk assessment matrix tables, Analytical Hierarchy Process (AHP), consequence assessment and cause consequence diagrams, Expected Monetary Value (EMV) using the Delphi technique, and influence diagrams are normally considered qualitative techniques.

Quantitative analysis relies on statistical methods and databases that determine the probability and consequence of an event. Simulation, Failure

Modes and Effects Analysis (FMEA), Fault Tree Analysis (FTA), Event Tree Analysis (ETA), success tree, accident progression and frequency analysis, common cause scenarios, sensitivity factors, fuzzy stochastic applications, risk premium, EMV and expected Net Present Value (NPV), risk adjusted rate of return, and stochastic dominance are generally considered quantitative risk assessment techniques.

Risk Management

Risk management is the process by which system operators, project managers, and owners make decisions, changes, and choose different system configurations based on the data generated in the risk assessment. Risk management is also dynamic as new information about risk events becomes available managers should adjust accordingly.

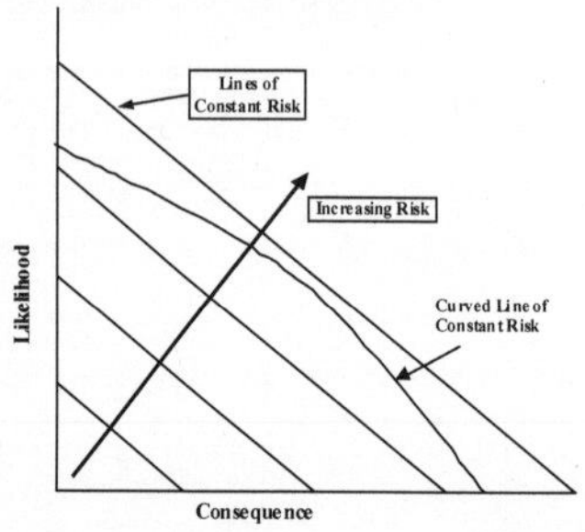

Fig. Risk Profile and Levels of Risk

In order to make decisions based on risk, a level of acceptable risk must be determined.

The lines of constant risk show that risk increases as the likelihood and/ or consequence of a risk event increases. When considering risk acceptance, curves that show a high consequence with an extremely low likelihood and vice versa may be used to show acceptable risk.

Management should determine risk acceptance through a systematic process that may be project specific, based on general corporate, or governmental guidelines. Risk acceptance can also be determined by the cost effectiveness of risk reduction or opportunity gained. This cost effectiveness is calculated as:

$$\text{Cost Effectiveness} = \frac{\Delta \text{Risk}}{\Delta \text{Cost}}$$

where DCost is the monetary amount required to reduce risk and DRisk is the level of risk reduction.

An integral part of risk management is thc use of decision analysis. Project management and engineering are professions that require decisions to be made for the management of cost, schedule technical, and safety risks. Most decision analysis techniques have the following steps or phases: 1) identify the problem and objectives, 2) develop alternatives, 3) evaluate the alternatives, and 4) implement the best alternative. In a risk-based decision analysis these steps should include the uncertainties associated with the data or alternatives. Four possible decision analysis methods that can be used in a risk based decision analysis methodology are:

1) decision trees, 2) goal trees, 3) Analytic Hierarchy Process (AHP), and 4) risk-based Net Present Value (NPV).

RISK-BASED COST CONTROL

A cost control technique that anticipates potential cost issues by using risk analysis and simulation techniques to highlight potential areas prone to cost escalation is presented. This risk analysis is also used to highlight areas where cost savings or a competitive advantage may be gained. The risk analysis techniques employed are a dynamic process, constantly updated, as new information becomes available. Risk analysis information along with cost forecasting tools is combined to anticipate a project's cost problems and completion costs.The presented methodology is applied in the planning and execution phase of a project. Risk methods are used to help establish cost and schedule targets in the planning phase. A project's costs are a function of schedule and the integration of cost and schedule is an important part of the methodology. In the execution phase risk methods are applied to help managers control projects with the goal of delivering them on budget and schedule.

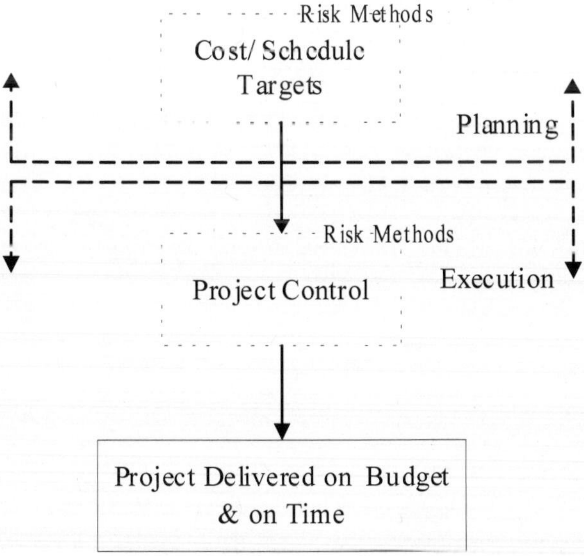

Fig. Risk-based Methodology Applied to both the Planing and Execution Phase

Table . Negative Risk Assessment Matrix Table

Likelihood level (1)	Consequence Assessment I Negligible (2)	II Acceptable (3)	III Marginal (4)	IV Critical (5)	V Catastrophic (6)
A. Implausible	N	L	L	L	M
B. Unlikely	L	L	L	M	H
C. Likely	L	L	M	H	H
D. Highly Possible	L	L	M	H	H
E. Certainty	L	L	M	H	H
Risk Assessment Guide					
N = Essentially no risk, can assume risk will not occur.					
L = Low risk, minor project cost escalation.					
M = Medium risk, average project cost escalation					
H = High risk, certain or if occurs will result in significant cost escalation.					

The presented risk analysis is a combination of both qualitative and quantitative risk methods. The generic structure of this risk analysis for cost control. In the risk assessment phase the probabilities and consequences of risks or opportunities are quantified and using a risk assessment matrix table a determination of risk is made.

An example of a risk assessment matrix table. Once an expression for the likelihood and consequence of an event is developed a risk rating can be determined. For example, if the likelihood of a labour strike is likely and the consequence of this event is critical, the risk rating for a labour strike is "high". Risk ratings are then compared to risk acceptance levels and a decision analysis process is used to assist management.

The activities in the boxes represent risk assessments and the activities in the arrow represent a shift to risk management functions. The basic structure is applied in both the planning and execution phases of a project except in the execution phase earned value techniques and continuous monitoring are performed.

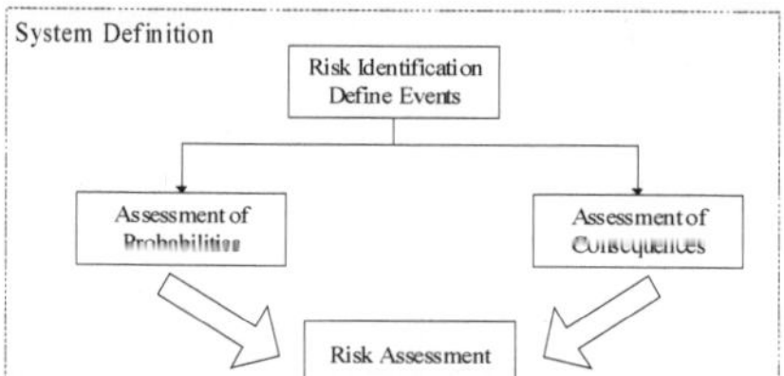

Fig. Generic System Definition of Risk-based Cost Control

PLANNING PHASE

A method to account for the uncertainty in estimated costs and schedule applies the results of a qualitative risk assessment method to simulation during the planning (feasibility or preliminary design) phase of a project to provide a qualitative and quantitative risk assessment. The qualitative method of the risk assessment is performed using a two-step process. First a check list approach is used to identify potential risk and then a risk assessment matrix technique is applied using expert judgement to quantify the risk. By using two methods of risk assessment potential areas of risk that affect the cost and schedule are accounted for.

The assessment of cost and schedule risks are then developed by a quantitative analysis using simulation.

Planning Cost and Schedule Risk Assessment

Construction project costs and schedules are typically estimated by developing a single value or "point estimate". A point estimate does not include the effects of uncertainty and is simply based on the summation of a number of point estimates for items of work.

A risk assessment methodology utilising simulation is presented because simulation techniques are readily adapted to the construction industry. This is because construction projects have multiple and interrelated activities that can be easily modeled.

Once a point estimate is developed, uncertainty can be probabilistically expressed by using probability density functions. Finally, simulation can be easily performed on current state of the art desktop computers.

One difficulty in using simulation to develop cost and schedule risks is determining what factors, parameters, or range of values should be applied to the probability distributions used to represent the uncertainty of costs and schedules.

The simulation model can then use information from the risk assessment to develop appropriate cost ranges modeled by the probability distributions.

The process for developing the cost and schedule risk data. The first step in establishing a cost estimate is to breakdown the project using a Work Breakdown Structure (WBS). Once this is done a point estimate can be developed for each individual work package.

Replacing a point estimate for a work package with a probability density function can better approximate the cost of each work package by accounting for uncertainty. Care will be required to use the appropriate density function and associated parameters.

The selection of which probability density function is best suited for certain construction operations has been demonstrated by AbouRizk and Halpin and Law and Kelton recommend several alternative distributions. The estimated cost and schedule target for a project is shown in the last block has been quantified to account for uncertainty and potential risks. This process also highlights risks or opportunities that may be mitigated or taken advantage of during the planning process.

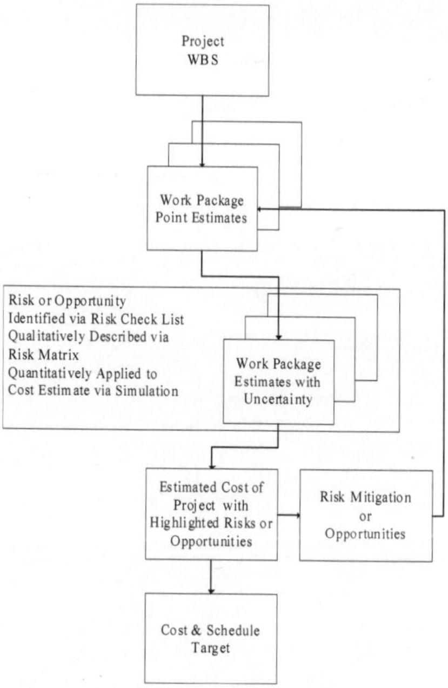

Fig. Cost and Schedule Risk Data and Process

Planning Cost and Schedule Risk Acceptability

Once the potential risks are highlighted from the risk assessment matrix a method of risk acceptability is required. The project team should propose to management the level of acceptable risk. The method for establishing acceptable risk should be based on criteria and guidelines established for a particular company or organisation. These criteria should be relatively generic so that they can be applied to projects typical for the organisation.

Risks that have the potential to cause significant cost impacts should be targeted for reduction or control. A two step process is used for determining risk acceptability: 1) all risk above a certain qualitative threshold should be targeted for reduction or opportunity, 2) all other risks should be evaluated by the cost effectiveness of risk reduction.

Management needs to set the thresholds for risk levels and cost effectiveness. Risk events with an acceptable cost effectiveness are deemed acceptable and mitigation efforts or opportunities will be pursued. Negative risks that are deemed unacceptable and not cost effective will likely make a project non-viable.

Planning Cost and Schedule Decision Analysis Method

There are three major decisions to be made in the planning process. The first one answers the question, "What risk should be mitigated or opportunities pursued in the planning phase?" The second decision area helps to set up the simulation for the development of cost and schedule targets. Third, a decision is required to determine if a project should be pursued furthered into the execution phase.

The output from the risk-based cost and schedule development also includes highlighted risk or opportunities. High negative risks and opportunities should be acted on during the planning phase. This should have the effect of reducing cost and schedule and therefore the original estimate must be redone. Finally cost and schedule targets are developed that include the effects of mitigating high negative risk and seizing opportunities.

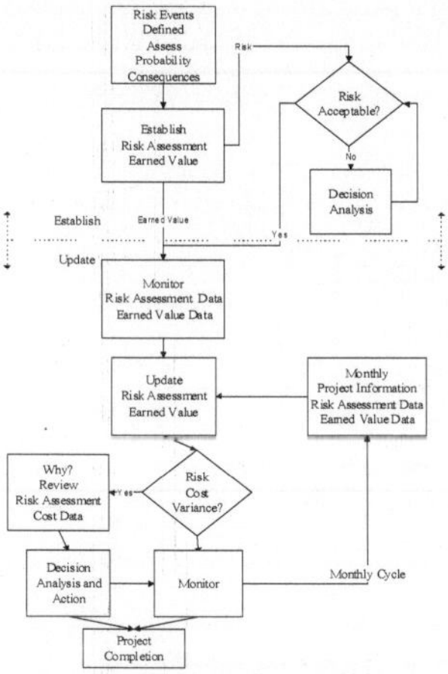

Fig. Combined Risk Assessment and Earned Value Analysis

The results of the risk assessment from the risk matrix are applied to the cost or schedule estimate via a simulation method. The risk assessment provides information for the determination of the appropriate range to use in the probabilistic representation of various cost or activities. The range is the measure of dispersion of a distribution about a mean value. For example, if a construction operation is represented by a triangular distribution the range of values may be arbitrarily selected from 90per cent to 110per cent of the mean. Though the results of the risk assessment this range can better represent the risk involved and is increased or decreased accordingly. For example, an activity with a high negative risk might use a range between 90per cent and 150per cent to reflect the potential for an adverse consequence.

The decision to go further with the project will be made once the risks have been highlighted and final target costs and schedules are known. Generally the most important risk issues are costs and schedule, the capacity of industry to perform the work, and the risks involved.

A methodology for using a risk-based Net Present Value (NPV) is used to determine if the project should be pursued further. This method is selected because it is relatively simple to apply, yields a monetary value that includes the time value of money, and can account for the uncertainty in a project.

THE DESIGN AND CONSTRUCTION OF FACILITIES FOR HANDLING CATTLE

Well-designed facilities for veterinary work, loading trucks, sorting and other procedures will make handling more efficient and help reduce stress and injuries. Reducing stress during handling is important because handling stresses can lower conception rates, suppress immune function, and raise cortisol levels. Rough handling will reduce weight gains and increase shrink. Cattle handled quietly in well designed facilities had much lower heart rates compared to cattle handled roughly in poor facilities.

The amount of stress imposed upon an animal during handling is greatly affected by its previous experiences. Cattle which have been handled gently will be quieter and less agitated when they are handled in the future. Weaner calves accustomed to regular gentle handling usually have less bruises during marketing because they are accustomed to handling. Animals remember aversive handling experiences for at least a year. Good facilities will reduce bruises and carcass damage on cattle and injuries to people. In the US, the cattle industry loses US$22 million annually from bruises (Livestock Conservation Institute, no date).

Bruise losses in New Zealand equal to 1per cent of the country's annual export beef earnings. Labour efficiency is greatly improved in good facilities. One person can provide a continuous flow of cattle into the squeeze chute for vaccinations, ear tagging and other procedures.

CATTLE PERCEPTION

Design of efficient handling facilities will be aided by an understanding of the behavioral characteristics of livestock. Cattle have 360' wide angle vision. They can see behind themselves without turning their heads and are sensitive to harsh contrasts of light and dark in loading ramps, races and handling areas. To facilitate cattle movement, illumination should be even, and there should be no sudden changes in floor level or texture. Even though ruminant animals have depth perception, their ability to perceive depth at ground level while moving with their heads up is probably poor. Hutson 0985b) suggests that there may be an extensive blind area at ground level and moving livestock may not be able to use motion parallax or retinal disparity cues to perceive depth. To see depth on the ground, the animal would have to lower its head. This would explain why cattle stop and balk at shadows. Cattle are more sensitive to high-pitch noises than people.

The sound of banging metal can cause balking and agitation. Rubber stops on gates and squeeze chutes will help reduce noise. The pump and motor on a hydraulic squeeze chute (crush) should bc locatcd away from the squeeze. On pneumatically powered equipment, silencing devices must be installed. The sound of hissing air will agitate cattle.

How to prevent balking

A single shadow that falls across an alley or race can cause balking. The lead animal will often stop and refuse to cross the shadow. Cattle will also balk at puddles of water, drain grates and bright spots of sunlight. Drains should be placed outside of races and crowd pens. Handlers should be cautious about causing moving shadows. Cattle have a tendency to approach a more brightly illuminated area, provided the light is not glaring in their eyes. Lamps directed towards the interior of a truck will facilitate loading at night. However, squeeze chutes and loading ramps should not facing the sun because cattle will not approach blinding light.

Sometimes it is difficult to drive cattle under a roof or into a building for handling. The animals will enter more readily if they are lined up in single file in a race. When a squeeze chute is inside a building or under a shade, the single race should extend at least 3-5 m outside the shade. Never place the edge of the shade or a building wall at the junction between the single-file race and the crowd pen.Cattle will also balk at moving or flapping objects. A coat flung over a fence or a shiny reflection off a truck bumper may stop the movement of cattle. If the cattle see people standing in front of the squeeze chute, they will refuse to approach. Installation of shields for handlers to stand behind may improve cattle movement. It will be easier to observe the distractions that are causing balking when the cattle are calm. Problems with balking tend to come in bunches. When one animal balks, the tendency to balk spreads to the next animal in line. An animal must never be prodded until it has an opening to move into. Cattle can be easily moved in large pens with a piece of cloth or a plastic tied to a stick. The animals will move away from the rustling plastic or the flapping cloth. Dogs should only be used in open areas where there is sufficient space for the cattle to move away. When dogs bite livestock, it is highly stressful.

Solid fences

The sides of the single-file race, loading ramp, and crowd pen should be solid. The crowd gate should also be solid to prevent cattle from attempting to turn back and rejoin their herd mates. The principle of solid fences is like putting blinkers on a harness horse. The solid fences prevent the cattle from seeing people, vehicles and other distractions outside the fence with their wide-angle vision. Solid race sides will help prevent wild cattle from becoming highly agitated in a race. Observations in a race with a solid and an open-sided portion indicated that some wild cattle are much more agitated in the portion of the race where they could see out. The cattle should see only one pathway of escape, up the single-file race. They will balk if the race entrance appears to be a dead end. Sliding or one-way gates at the junction between the single-file race and the crowd pen must be constructed from bars or mesh so that cattle can see through them. However, a gate which is used in a single-file race for stopping cattle movement during sorting or dipping should be solid to prevent excited animals from attempting to push through it. All other races and forcing pen fences should be solid.

On steel corrals, it would be too expensive to construct all the holding pens, sorting pens and alleys with completely solid fences. On fences built from pipe or rod, a 30-60 cm-wide belly rail placed at cow eye height will facilitate movement and will prevent the ramming of the fence by excited cattle. This is especially important in facilities where wild Brahman, Brahman cross and Zebu cattle are handled because Brahman-type cattle are more excitable and difficult to block at gates. A belly rail is also recommended for handling excitable genetic lines of European continental cattle. Corrals constructed from wide wood planks do not need an additional belly rail because the boards create a substantial visual barrier.

Flight zone

When a person enters an animal's flight zone it will move away. If the handler penetrates the flight zone too deeply, the animal will either bolt and run away or turn back and run past the person. When the flight zone of a group of bulls was invaded by a mechanical trolley, the bulls moved away and maintained a constant distance between themselves and the trolley. The best place for the person to work is on the edge of the flight zone. This will cause the cattle to move away in an orderly manner. The animals will stop moving when the handler retreats from the flight zone. To make an animal move forward, the handler must be positioned behind the point of balance at the shoulder.

Fig. Well-designed curved single-file race with solid sides and a walkway along the inner radius for the handler.

The size of the flight zone varies depending on the tameness or wildness of the cattle. The flight zone for extensively raised cows may be as much as 50 m, whereas the flight zone of feedlot cattle may be 2-8 m. The edge of the flight zone can be determined by slowly walking up to the animals. The circle represents the edge of the flight zone. Extremely tame cattle are often difficult to drive because they no longer have a flight zone. The size of the enclosure in which the livestock are confined in may affect the size of the flight zone. Sheep experiments indicated that animals confined in a narrow alley had a smaller flight zone compared to animals confined in a wider alley. Flight distance is also affected by previous experience. Cattle with previous experiences with gentle handling will have a smaller flight distance than cattle which have been handled roughly, and animals raised intensively in buildings in close contact with people will have a smaller flight distance than extensively raised range cattle.

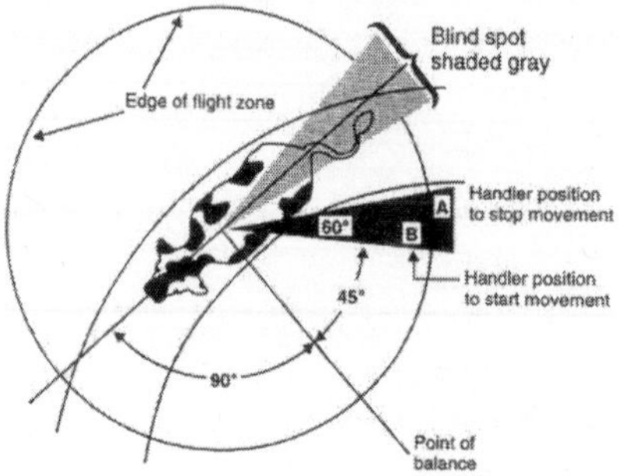

Fig. Correct handler positions for driving cattle. The handler should remain on the boundary of the flight zone for most efficient cattle movement. He moves from position A to position B to make the animal move forward.

Many people make the mistake of deeply invading the flight zone when cattle are being driven down an alley or into an enclosed area such as a crowd pen. If the cattle attempt to turn back, the person should back up and retreat from inside the flight zone. The cattle attempt to turn back because they are trying to escape from the person inside their flight zones. Cattle sometimes rear up and become agitated while waiting in a single-file race. A common cause of this problem is a person leaning over the race and deeply penetrating the flight zone. The animal will usually settle back down if the person backs up and retreats from the flight zone.

Curved race more efficient

A curved race is more efficient than a straight race for two reasons. First, it prevents the cattle from seeing the truck or squeeze chute until they are almost in it. A curved race also takes advantage of the animal's tendency to circle around the handler. Cattle will face a handler who enters their pen. As the handler moves through the pen, the animals will circle around him. A curved race takes advantage of the natural tendency to circle around a person. A curved race provides the greatest advantage when the cattle have to wait in line for vaccination or other procedures. Experiments with continuously moving livestock indicated no significant difference between straight and curved races for run-through time. However, when a curved race with a round crowd pen is used in a practical situation with the livestock lined up and waiting to enter a squeeze chute, the curved system is faster. Circular crowding pens and curved races can reduce the time spent in moving cattle by up to 50per cent.

Cattle can be driven most efficiently if the handler is situated at a 45-60' angle to the animal's shoulder. A well designed curved race has a walkway for the handler along the inner radius. The handler is forced to stand in the correct position to facilitate animal movement. The curved lines represent the curved race. The solid fences block out all visual distractions except for the handler on the catwalk.

In Australia and South America, cattle are often given veterinary treatment in the single-file race, whereas in the US, Canada and many European countries, the animals are treated while held in a squeeze chute (crush) or head gate (stanchion) at the end of the race. In a South American or Australian operation, completely solid race sides would block access to the cattle. In this situation, the outer fence should be completely solid and the fence on the inside radius should be constructed from pipe or wood planks with spaces between them. The handler walkway is omitted.

To prevent leg injuries, the inner radius fence should have a 60 cm high solid panel at the bottom. In curved race systems with completely solid sides, the handler walkway should run alongside the race and never be placed overhead. The distance from the walkway platform to the top of the race fence should be 100 cm.

Curved race and crowd pen dimensions

For feedlot and range cattle handling facilities, the recommended inside radius for a curved race is 3.5-6 m. A single-file race must be long enough to take advantage of cattle following behaviour. The minimum length for a race used for handling large numbers of cattle is 9 m. A half circle race with a 3.5-5 m inner radius is the ideal length. Excessively long races are not recommended because some cattle have a tendency to lie down and get trampled if they are held too long in a race.

The longer race with the 6 m inside radius is recommended when cattle are vaccinated rapidly while held in the race. When a race is designed, care must be taken to avoid bending the race sharply at the junction between the single-file race and the crowd pen. A sharp bend at this point will make the entrance to the race appear to be a dead end. The cattle will balk and may refuse to enter. An animal standing in the crowd pen must be able to see a minimum of two body lengths up the single-file race.

Curved races can be built from wood, steel or concrete. When wood or concrete is used, the race can be built in a series of straight sections. The posts should be spaced 1.2 m apart. To reduce construction costs, the race has a single board for a person to step on to prod cattle. For large numbers of cattle, a complete handler walkway should be constructed.

If space is restricted, a race with an inner radius as small as 1.5 m can be used if certain rules are followed. When the inner radius is shorter than 3 m, the race must have a minimum of a 3 m-long straight section joining the race to the crowd pen. This prevents the race from appearing to be a dead end. A race with a very short inner radius must be built in a continuous smooth curve. Cattle will get stuck if it is built in a series of straight sections. The recommended race width for a race with straight sides is 66-71 cm for adult cows and 51 cm for calves.

Fig. Curved wood, wide curved lane with solid sides which leads to the round forcing pen.

These dimensions may vary depending on cow size. A V-shaped race should be 41-45 cm wide at the bottom and 81 cm wide at the 152 cm level. A common mistake is to make the race too wide.

There should only be 2 cm of clearance on each side of the largest cow which will use the race. Fence height for races, crowd pens and corrals is 152 cm for English breeds and tame cattle, and 167-183 cm for Brahman, Brahman cross, Zebu excitable genetic lines of European continental or wild cattle.

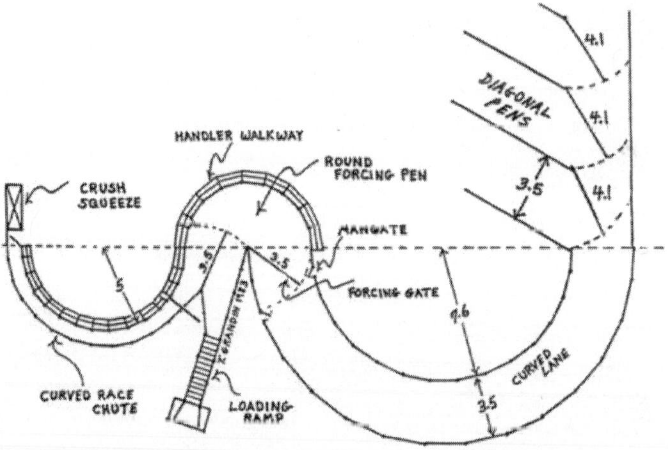

Fig. Basic cattle handling system with curved race, round crowd pen and curved lane.

A curved race, round crowd (forcing) pen and a wide curved lane. The crowd pen has solid fences and is equipped with a crowd gate which can be advanced behind the cattle.

This gate should be solid. The wide curved lane holds cattle which are waiting to go into the crowd pen. A single person can easily move them into the crowd pen by working along the inner radius of the 3.5 m wide lane. This basic layout can be used on ranches, in feedlots for the main cattle handling facility, and in slaughter plants.

For small handling facilities on farms or feedlot hospitals, the wide curved lane can be deleted and the round crowd pen can be connected directly to existing alleys. The designed for easy layout. It consists of three half circles which are located along the dotted line. If the crowd pen is constructed from wood, the fence can be built in a series of straight sections with 1.2 m post-spacing. The ideal length for the crowd gate is 3.5 m. A longer gate is unwieldy and a larger crowd pen is inefficient because it holds too many cattle. For smaller operations, the gate length can be reduced to 3 m. A well designed round crowd pen with solid fences, solid crowd gate and a walkway for the handler. The handler can advance the crowd gate as he walks along the walkway.

This crowd pen can be used to direct cattle into a single-file race or into a single-file loading ramp. There is one mistake; the sliding gate at the junction between the single-file race and the crowd pen should be constructed from bars so that the cattle can see through it.

A solid sliding gate makes the race entrance look like a dead end. Research indicates that solid fences for crowd pens are more efficient. Cattle moved faster through a crowd pen with solid fences compared to one with a pipe fence or partially open-board fence. The inside is smooth to prevent bruises. All structural supports are on the outside.

If a straight crowd pen is used with a funnel leading to the single-file race, one side should be straight and the other side should be on a 30' angle (Meat and Livestock Commission, no date). If space permits, a round crowd pen is recommended because it is more efficient, as some types of cattle move more slowly when they walk straight through the crowd pen into the single-file race.

Mangates should be installed as indicated, and they should be 45 cm wide because large cattle cannot pass through the narrow opening. A hinged solid metal or plywood flap which opens inward towards the cattle makes a good mangate. The flap is held shut with a spring and there is no latch. If a person is chased by the cattle, he can open the gate quickly because there is no latch.

Loading ramps

A gate which can be used to direct cattle to either the loading ramp or the single-file race to the squeeze. In the US, the most efficient loading ramps are single-file, because US trucks have a narrow 76 cm wide rear door. Therefore, the ramp should be 76 cm wide for adult cows and fattened cattle. This is narrow enough to prevent adult cattle from turning around. If the ramp is used for calves only, it should be made narrower.

The efficiency of the ramp can be further improved by curving the single file ramp. The use of a ramp wider than the truck door is not recommended for loading, because it is inefficient and the cattle will become bruised when they strike the door frame. In countries where the back gate of the truck opens up to the full width of the truck, a ramp equal to the truck width can be used. In the US and other countries where trucks with narrow doors are used, a 2.5-3 m-wide ramp is recommended for unloading only.

Many animals are injured on loading ramps which are too steep. The maximum recommended angle is 20' for permanent ramps and 25' for adjustable ramps. The crowd pen on a single-file loading ramp must have a level floor except for a slight drainage slope. Sloping the floor of the crowd pen 10' will cause livestock to pile up against the crowd gate. On concrete ramps, stair steps are recommended.

The dimensions for the stair steps are a 30-cm tread width and a 10-cm rise. The steps should be deeply grooved to provide a Non-slip surface. On wooden ramps, the cleats should have 20 cm of space in between them. To help prevent falling during unloading, permanently installed ramps should have a flat-level dock at the top.

The minimum width for the level dock is 1.5 m. A self-aligning dock bumper will help prevent injuries caused by cattle stepping down between the truck and the loading-dock. Even if the truck backs up to the dock into a misaligned position, the gap is blocked by the self-aligning bumper. The bumper can be constructed from two pieces of steel welded together to form an 'L' shape. The 'L'-shaped piece of metal pivots on a heavy steel pin attached to the front of the dock. One side of the 'L' overlaps and rests on the dock floor. Loading ramps should also have telescoping side panels to prevent cattle from jumping out between the truck and the ramp. If a portable ramp is used, it should be sturdy. Ramps which sway or move when cattle walk on them are likely to cause balking.

Working corral for a large ranch

A large corral system for gathering cattle for transport, sorting, working through the squeeze chute, weighing and other chores. It can be used by handlers on foot or people mounted on horses. It is especially suited for ranches where calves or yearlings are removed from the ranch annually for shipment to feedlots.

The layout forms the left-hand side of the corral. The only difference is that the inside radius of the wide curved lane is expanded to 10.66 m. Most of the actual cattle handling and sorting is conducted in the wide curved lane labeled 'sorting reservoir', the single-file race, round crowd pen and diagonal pens. Cattle are more easily controlled in the 3.5 m-wide lanes and pens. The large gathering and post-working pens are only used to hold cattle before and after the actual handling operation. The curved lanes and diagonal pens eliminate square comers and promote cattle movement. The gates on the diagonal pens are 4.2 m long on a 3.5 m-wide lane because gates opening on an angle eliminate sharp comers. The round gathering and holding pens have no square comers for cattle to bunch up in. To set out by placing strings on the ground as indicated by the dotted lines. To prevent mistakes, the entire corral should be set out and the ground should be marked with lime before starting construction.

Fig. Corral constructed from wood planks.

Groups of 20-40 animals are directed from the gathering pen into the curved sorting reservoir lane. This lane serves two functions. First, it holds cattle which are waiting to go to the loading ramp, or the squeeze chute, in the same manner. Secondly, it holds groups of cattle which are being sorted back into the diagonal pens. Sorting back into the diagonal pens is efficient because the animals have a strong tendency to move back in the same direction from which they came. Many US ranchers prefer sorting back into the pens from an alley because it is quick, and it enables them to see the animals more easily than sorting through a single-file race. Cattle which have been sorted into the diagonal pens can either be released back to pasture or moved into the curved lane to go to the scale, loading ramp or squeeze chute.

When cows and calves are being separated, the calves are sorted into the diagonal pens. The cows are allowed to pass through one of the diagonal pens into the large postworking pen. This corral system can handle 300 cow-and-calf pairs or 400 adult cattle. To expand the systems, additional diagonal pens can be added. The length of the diagonal pens should not be increased. If they are too long, the cattle will bunch up. Increasing the size of the gathering pen is not recommended. If it is too big, driving cattle into the curved lane may be difficult. To increase the gathering area, an additional round gathering pen should be built at the pasture entrance. The corrals can be reduced for smaller herds by omitting one or two of the diagonal sorting pens and reducing the size of the gathering and post-working pens. The basic round shape of the pens should be maintained to eliminate comers for cattle to bunch up in.

A sorting gate in front of the squeeze chute. When cows are pregnancy-tested, the pregnant cows can be directed to the post-working pen, and the Non-pregnant cows can be directed into one of the diagonal pens. A second sorting gate and alley can be easily added to create a three-way sort out of the squeeze chute. The sorting gate or gates in front of the squeeze chute can also be used for high-speed sorting as cattle walk through the squeeze. Any animal which needs veterinary treatment can be easily caught in the squeeze. An added advantage of sorting through the squeeze chute is that the cattle will learn to enter it readily. As an added incentive to enter the squeeze, feed can be made available in the postworking pen. Feeding palatable barley grain to sheep immediately after handling reduced the time required to drive them through a race.

This corral can also be used in pasture rotation systems which have centrally located handling facilities and the pastures are laid out like a wagon wheel. The gathering pen and post-working pen are eliminated and replaced with a 6 m-wide lane which encircles the corral and forms the hub of the wheel. Pasture fences radiate from the 6 m-wide lane. Switching cattle from pasture to pasture is easy when they come in for water in the 6 m-wide lane.

In feedlots by eliminating the gathering and post-working pens and connecting the lanes to the alleys in the feedlot. Additional feedlot handling system layouts are available from Paine et a]. (no date). Paine's publication contains layouts of diagonal pens for shipping, receiving, weighing and loading cattle in large feedlots, and it also discusses feedlot hospital design.

The corrals are most suitable for use with British, European or British/ European crosses with Brahman. These types of cattle can be readily sorted back into the diagonal pens by cutting animals out of the reservoir lane one at a time. Purebred Brahmans and Zebu tend to mill and circle more tightly. Cutting out animals and sorting them back is often more difficult.

For large properties (ranches) in Australia or South America which handle Brahman or Zebu cattle fattened on grass. In this type of operation, the steers remain on the ranch for several years instead of being shipped to a feedlot. More sorting is required due to the greater range of cattle ages and types. Older steers which have been sorted many times may be harder to sort back in an alley than inexperienced calves and yearlings, hence the Australians developed the pound yard. It enables the person sorting the cattle to look at each animal carefully before a sorting decision is made. However, it is slower than sorting back in the alley.

The advantages of both high speed sorting in a single-file race and pound yard sorting. Cattle can be sorted three ways with the two sorting gates in the single-file race. During sorting, all cattle have to pass through the headgate and squeeze chute.

They can be sorted five ways out of the 5-m diameter pound yard. Each animal is admitted one at a time. A person on a platform over the pound yard can easily open and shut the gates with ropes or levers. A triangular block gate is used to stop incoming cattle and control cattle flow into the pound yard. It consists of 1.52 in high solid sided triangle with 76cm sides which is hinged at its apex. After sorting, the cattle in the sorting pens can be easily moved through the return lane for vaccinating, branding, truck loading, etc. If a dip vat is required, an additional directional gate and race can be added to lead to the dip vat. Easy to set out and build by using the dotted lines as a guide. Sorting pens 2 through 6 are in a half circle with a 17.8-m radius. For electronic sorting of cattle. Producers in the US are now doing more individual animal evaluation and there will be an increasing need for the type of layout.

SQUEEZE CHUTES AND HEADGATES

A good headgate and squeeze chute will improve the care and management of cattle health because catching and restraining cattle is easy. There are many different types of commercially available headgates for restraining the animal's head. Headgates can also be built from plans available from Midwest Plan Service, 1975, Inglis and Williams, 1979 and Vowles, 1980. The four basic types are scissors stanchion, full opening stanchion, positive control and self catcher.

Scissors stanchion

It consists of two biparting halves that have pivots at the bottom. After release, the animal walks out through the headgate. It is available in the curved bar type or a straight bar. The curved bar stanchion is one of the most popular general-purpose headgates. The curved bar provides better head control because it prevents. Squeeze chute with scissors stanchion headgate with curved neck bars. animal from sliding its head up and down. The animal may choke if it lays down in the chute. The straight bar provides poor head control because the animal can slide its head up and down. Choking in a straight bar stanchion is almost impossible because the straight bars can not press on the throat. A curved bar stanchion is recommended for general cattle handling on feedlots and ranches. A straight bar stanchion is recommended for gentle dairy cows and for veterinary clinics where an animal must remain in the headgate for a long period. It is also recommended if the primary use of the headgate is restraining cows for pregnancy checking or artificial insemination or when a headgate is used alone without a squeeze chute.

Full opening stanchion

This consists of two biparting halves which open and close like a pair of sliding doors. It is available in both straight bar and curved bar models. The advantage of this type of headgate is that large bulls can walk through it more easily. The disadvantage is that the sliding mechanism is more complicated.

Self-catching

This headgate can be set like a trap to automatically catch the animal's head when it enters. Forward movement of the animal will close the gate around its neck. Self-catching gates are recommended for gentle cattle without horns. To prevent injuries to the cattle and damage to the gate, the cattle should walk slowly into the headgate. The mechanism is complex and needs constant adjustment. These gates work best on small ranches or dairies where a single person handles gentle cattle. They are available in both curved and straight bar stanchions.

Positive control

This headgate locks very tightly around the animal's neck like a pillory. It provides excellent control of the head, but it is more likely to choke animals than a curved bar stanchion. It is recommended for wild cattle with homes because it is easier to catch homed cattle with this type of gate. After the neck is released, the animal must back up before the gate is swung open to allow it to exit. Another advantage is that it requires less effort to operate than the other types of head gates.

Choking in a headgate is usually asphyxiation caused by excessive pressure on the carotid arteries in the neck or on the wind-pipe. Due to the pressure on the arteries, animal can die very rapidly if it starts to lose consciousness in a headgate. The headgate must be released instantly when the first signs of asphyxiation occur. Choking is most likely to occur when a headgate is used without a squeeze. A properly adjusted squeeze chute can greatly reduce choking by preventing the animal from lying down. The best squeeze chutes have two movable side panels which are hinged at the bottom and pulled together by a lever system at the top. These are superior to chutes with a single movable side because the animal remains standing in a balanced position. The 'V' shape of the squeeze sides supports the animal. Proper adjustment of the space between the squeeze sides at the floor can greatly reduce choking. For 113-180 kg calves, the squeeze sides should be 16 cm apart at the chute floor, 21 cm for 272-360 kg cattle and 30 cm apart for most cows and fed steers. For large bulls the spacing may need to be wider.

On commercially available squeeze chutes, the sides have bars which can be dropped down for access to the sides of the animal. The solid panel at the bottom can also be opened for access to the underside of the animal. When a squeeze chute is being purchased, the position of the control levers should be considered. On some headgates and squeeze chutes, the levers are situated where they may injure the operator if a latch is accidentally released. Commercially built headgates and squeeze chutes have two basic types of latches. The first type is a ratchet-latch which locks into a definite notch as the headgate or squeeze is closed. It has the disadvantage of being noisy, but it is safer because it is less likely to be released accidentally. The second type is a friction latch which consists of a steel rod which passes through a hinged metal plate. It has the advantage of being quieter than a ratchet latch, but it is more likely to come unlatched accidentally. Friction latches must be well maintained to keep them safe.A survey conducted in large feedlots by Grandin (1980b) indicated that operator carelessness and trying to handle cattle too rapidly was the primary cause of choking, escaping and legs caught in squeeze chutes. The survey results also indicated that Brahman cross cattle were more likely to escape from a squeeze chute than English/European cross cattle. Allowing cattle to run rapidly into a squeeze chute and slam against the headgate can cause serious injuries. Examination of beef carcasses revealed old, healed spinal injuries in the back and neck. Even animals which appeared to be normal may have had hidden spinal damage.

A skillful squeeze-chute operator can slow cattle down before they reach the headgate by partially closing the squeeze. Injuries can also be reduced by handling cattle quietly in the race leading up to the squeeze chute. Excessive use of electric prods especially on Brahman, Brahman cross and Zebu cattle can increase squeeze-chute injuries because excited cattle slam into the headgate and make greater attempts to escape.

To prevent shoulder bruises, the headgate should have neck bars constructed from round pipe with a minimum diameter of 6.2 cm. A 7.6 cm diameter pipe is recommended. The larger pipe diameter is less likely to bruise the neck. Headgates can be padded with old conveyor belts or split tires. Split motorcycle tires are the ideal size for headgate stanchions.

Many large feedlots and some ranches use hydraulics instead of muscle power to operate the squeeze chute. A correctly adjusted hydraulic squeeze chute is usually safer for both people and animals. The dangerous protruding levers are eliminated, and people are less likely to become tired and make errors which can cause an accident.

Most commercially available hydraulic squeeze chutes in the US have a factory adjusted pressure relief valve which prevents excessive pressure from being applied to the animal. Cattle can be seriously injured if excessive squeeze pressure is applied. Animals which have been oversqueezed will sometimes appear to have pneumonia symptoms a few days later. Autopsies of cattle which have died from oversqueezing indicated that they had internal ruptures. When a hydraulic squeeze chute is designed, the force exerted on the animal should be determined. Measurements with a hydraulic load cell of the force exerted by the squeeze sides indicated that the recommended force 69 cm from the bottom pivots at a single point is 454-680 kg for cattle weighing over 272 kg and 270362 kg for cattle weighing less than 272 kg. These force readings are not hydraulic system pressure. Oversqueezing is most likely to occur if the pump motor supplied with the squeeze chute is substituted with a larger motor or if a tractor hydraulic system is used. If an animal has difficulty breathing while held in a hydraulic squeeze chute, the pressure relief valve should be loosened.

CALF TABLES

Young calves on many US ranches are restrained for branding, castration and dehorning by roping them with a lariat. Roping calves properly so excitement is minimised is a highly skilled occupation. Many ranches now use a calf table to restrain calves. This is a miniature squeeze chute which can be tilted to the horizontal position. Commercially available calf tables are available with the four different types of headgates. Some calf tables have no headgate and the calf table is squeezed and tilted with the same lever. A well designed calf table requires little physical effort to bring it to the horizontal position. Tables are available which will handle up to 200 kg calves. For tilting adult cattle to the horizontal position for foot trimming, two basic types of equipment are commercially available: a tilting table on which the animal is secured by two wide belly straps, or a tilting squeeze chute. The tilting squeeze chute is safer for both the operator and the animal.

Artificial insemination chute

For improved conception rates, cows should be handled gently during artificial insemination. They should not be allowed to become agitated and overheated. The chute used for artificial insemination should not be used for painful procedures such as dehorning, branding or having her head pulled around and restrained with nose tongs. Nose tongs are very aversive to cattle and they will attempt to avoid them after having experienced them once. A less aversive form of head restraint is a rope halter.

Cows can be easily restrained for artificial insemination or pregnancy testing in a dark-box chute. It has no headgate or squeeze sides and it will hold the wildest cow with minimum excitement. The dark box is 66-71 cm wide depending on cow size, and consists of completely solid sides, solid front, and a solid top. A piece of cloth is hung over the cow's rump to make it completely dark. When the cow is inside the box, she is in a snug dark enclosure. If the cows refuse to enter the dark box, a small window can be cut in the front gate. Cow entry is usually not a problem if a good single-file race leads up to the dark box.

The dark box works on the same principle as the dark room which is used for handling deer in New Zealand. Groups of deer brought into the dark room will allow people to touch them and remain clam. Outside the dark room, the deer would become agitated and attempt to jump high fences. The dark box and the dark room may reduce physiological arousal levels. Experiments with poultry indicated that blind-folded birds had lower heartrates and respiration rates during shackling and slaughter. Preliminary experiments with cattle are yielding a similar result. A new novel dark box can cause stress. Prior to breeding cattle should be handled in the box so that they can become familiar with it.

If wild cows are to be handled, an extra-long dark box can be constructed. A tame cow which is not displaying estrus is placed in front of the cow which will be inseminated. A wild cow will usually stand quietly and place her head on the 'Pacifier' cow's rump. Cattle will often remain calmer when they are in bodily contact with other cattle. After insemination the cow is released through a side gate. The 'pacifier' cow remains in the dark box. If large numbers of cows are going to be pregnancy checked or inseminated, 26 dark-box chutes can be built side by side in a herringbone pattern on a 60' angle. They are built like regular dark-box chutes and the cows exit through the front of each chute. On some herringbone systems a single, large front gate is used to release all the cows at once. The outer fences, front gates, and tops are solid. The fences in between the cows are constructed from bars so the cows can see and feel each other. This will help keep them calmer.

NEW RESTRAINT IDEAS

There is a need to improve restraint devices for handling cattle, especially for extensively raised animals that are often wild. When untamed semi-wild cattle are handled in a squeeze chute, 1.6-7.8per cent are bruised. Most of these bruises are caused by hitting the headgate too hard. From a behavioral standpoint, existing squeeze chutes are poorly designed. The open barred sides permit cattle to see the operator who is deep inside the animal's flight zone. This may cause the animals to become agitated. The installation of solid sides and a solid barrier in front of the headgate to block the animal's vision will keep a semi wild animal calmer. The behavioral principles of restraint are: blocking the animal's vision; slow, steady motion of the equipment and optimal pressure. The device must apply sufficient pressure to provide the feeling of being held, but excessive pressure that causes pain and struggling should be avoided. Semi-wild cattle will remain calmer if restraint in a headgate is accompanied with body restraint. Slow steady motions of people and equipment are calming and sudden jerky motion excites and agitates cattle.

Design ideas from equipment used in slaughter plants should be adopted for ranch and feedlot use. A double rail conveyor restrainer used for beef cattle in slaughter plants outfitted with a head restraint device would almost eliminate injuries to cattle caused by lunging against the headgate. The animals straddle a moving conveyor.

DIPPING VATS

Pharmaceuticals are reducing the need for dipping cattle to eradicate external parasites. Ivermectin has replaced dipping on many cattle ranches and feedlots in the US. There is still a need for dip vats in areas where livestock have to be dipped frequently due to the high cost of ivermectin. Dipping is still required in some quarantine areas because Ivermectin does not kill all the parasites immediately. The eventual replacement of dip vats with injectable or pour on products is beneficial because dipping is stressful and the disposal of used dip chemicals may create pollution.

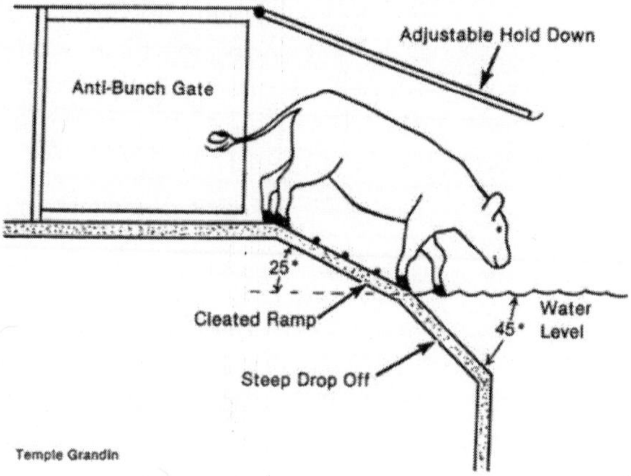

Fig. Dip vat entrance design which will reduce injuries and chemical splashing.

Injuries, stress and chemicals splashing on people can be reduced by a well designed entrance to the dip vat. Many injuries and drownings occur because too many cattle enter the dip vat at once or they jump on top of each other. The two anti-bunch gates can be adjusted to allow only one animal to pass through at a time. The pair of gates is located on each side of the single-file race. The opening between the ends of the two gates is adjusted for animal size. To prevent wild cattle leaping into the vat, an overhead rack directs the animal head first into the water.

Over 95per cent of the cattle will fully submerge their heads and will not have to be pushed under with a stick. The hold-down rack also reduces chemical splashing. To further reduce splashing, an 8 cm-diameter pipe should be installed on both sides of the vat wall approximately 1 m above the water. Splashing water will hit the pipe and fall back into the vat.

Each animal enters the vat by walking down a ramp which is deeply grooved to prevent the animal from slipping. The ramp is on a 20-25' angle. The animal can stand on the ramp without slipping. When it steps out over the steep drop off, its center of gravity will change and it will fall into the water. The animal will seldom attempt to back out because it does not start slipping. Many vat builders make the mistake of building a slide. A slide is a bad design because the cattle sometimes flip over backwards while going down the slide. More detailed information on vat design and construction can be found in Grandin (1980a,c, 1983a). For specific information on chemical use and disposal contact, it should be made with the local agricultural officials in each country. Universities and pharmaceutical companies can also provide information on dip vat management and chemical usage.

BRUISE AND INJURY PREVENTION

Careful, quiet handling will greatly reduce bruises. Fifty percent of all bruises are caused by rough handling. Surfaces which contact cattle should be smooth. Sharp comers should be padded with old conveyor belts or split tires. A smooth, flat surface such as the inside of a race does not need padding. Bruises are most likely to occur when animals hit an object with a small diameter such as the edge of a steel bar or a nail sticking out of a fence. An animal striking the comer of a square 10 cm diameter post is more likely to bruise than an animal striking a round 10 cm diameter pipe post. Gates should be equipped with tie-backs to hold them back against the fence. A gate swinging out into an alley can seriously bruise an animal if it becomes jammed between the end of the gate and the fence. The use of sticks, metal pipes and sharp objects for driving cattle should be forbidden. Guillotine gates which slide up and down should be counterweighted to prevent them from injuring an animal's back. The bottom of the guillotine gate should be constructed from a 7.5 cm diameter pipe to prevent bruises. If an air cylinder is used to actuate the gate it should be connected to the gate with a cable. This will prevent back injuries because gravity will close the counter balanced gate. Back injuries caused by a powerful cylinder forcing the gate down on an animal are prevented.

Animals can become crippled and injured if they slip and fall on slippery concrete floors. Cattle handling facilities should have Non-slip floors. In races, crowd pens, scales and other cattle-handling areas, concrete floors should be deeply grooved. The grooves should be made in both directions in a 20 cm square or diamond pattern. The grooves should be 2.5-5 cm deep. In existing facilities, concrete floors can be roughened with a pneumatic hammer or a grooving machine. These recommendations are for handling facilities where cattle are handled intermittently such as auctions, feedlots and ranches. The deep grooves described above should not be used in the animal's living quarters or in milking parlors where dairy cows walk twice a day. The deep grooves will cause excessive hoof wear in these locations.

WASHABLE FACILITIES

In large feedlots, veterinary clinics and dairies, handling facilities should be designed so the squeeze chute, single-file race and the concrete slab around the squeeze chute can be easily washed. Drains should be located outside of the areas where cattle will walk. Concrete floors should be sloped 0.630.30 cm every 30 cm towards a drain. Curbs should be installed to contain the wash water and direct it to a drain. The best type of drains are open concrete ditches. Square concrete ditches should be constructed slightly wider than the width of a shovel for easy cleaning. One good drain design is to locate a large drain directly under the squeeze chute and slope the floor towards it from all sides. Cattle can not see the drain under the squeeze-chute floor.

CONSTRUCTION MANAGEMENT

Construction management or construction project management (CPM) is the overall planning, coordination, and control of a project from beginning to completion. CPM is aimed at meeting a client's requirement in order to produce a functionally and financially viable project. The construction industry is composed of five sectors: residential, commercial, heavy civil, industrial, and environmental. A construction manager holds the same responsibilities and completes the same processes in each sector. All that separates a construction manager in one sector from one in another is the knowledge of the construction site. This may include different types of equipment, materials,subcontractors, and possibly locations.

In 2010 the Chartered Institute of Building published its construction management definition based upon the Institute's 45,000 members and the work they do. The designation Chartered Construction Manager is available to full corporate members (MCIOB) and fellows (FCIOB) of the Chartered Institute of Building.

THE ROLE OF A CONTRACTOR

A contractor is assigned to a construction project once the design has been completed by the person or is still in progress. This is done by going through a bidding process with different contractors. The contractor is selected by using one of three common selection methods: low-bid selection, best-value selection, or qualifications-based selection. this is the main role of a contractor.

A construction manager should have the ability to handle public safety, time management, cost management, quality management, decision making, mathematics, working drawings,and human resources.

FUNCTIONS

The functions of construction management typically include the following:

1. Specifying project objectives and plans including delineation of scope, budgeting, scheduling, setting performance requirements, and selecting project participants.
2. Maximizing the resource efficiency through procurement of Labour, materials and equipment.
3. Implementing various operations through proper coordination and control of planning, design, estimating, contracting and construction in the entire process.
4. Developing effective communications and mechanisms for resolving conflicts.:

The Construction Management Association of America (a US construction management certification and advocacy body) says the 120 most common responsibilities of a Construction Manager fall into the following 7 categories:

Project Management Planning, Cost Management, Time Management, Quality Management, Contract Administration, Safety Management, and CM Professional Practice. CM professional practice includes specific activities, such as defining the responsibilities and management structure of the project management team, organizing and leading by implementing project controls, defining roles and responsibilities, developing communication protocols, and identifying elements of project design and construction likely to give rise to disputes and claims.

SEVEN TYPES OF CONSTRUCTION

- Agricultural: Typically economical buildings, and other improvements, for agricultural purposes. Examples include barns, equipment and animal sheds, specialized fencing, storage silos and elevators, and water supply and drains such as wells, tanks, and ditches.
- Residential: Residential construction includes houses, apartments, townhouses, and other smaller, low-rise housing types.
- Commercial: This refers to construction for the needs of private commerce, trade, and services. Examples include office buildings, "big box" stores, shopping centers and malls, warehouses, banks, theaters, casinos, resorts, golf courses, and larger residential structures such as high-rise hotels and condominiums.
- Institutional: This category is for the needs of government and other public organizations. Examples include schools, fire and police stations, libraries, museums, dormitories, research buildings, hospitals, transportation terminals, some military facilities, and governmental buildings.
- Industrial: Buildings and other constructed items used for storage and product production, including chemical and power plants, steel mills, oil refineries and platforms, manufacturing plants, pipelines, and seaports.
- Heavy civil: The construction of transportation infrastructure such as roads, bridges, railroads, tunnels, airports, and fortified military facilities. Dams are also included, but most other water-related infrastructure is considered environmental.
- Environmental: Environmental construction was part of heavy civil, but is now separate, dealing with projects that improve the environment. Some examples are water and wastewater treatment plants, sanitary and storm sewers, solid waste management, and air pollution control.

CONSTRUCTION MANAGEMENT JOBS

- Project executive
- Project manager

- Planning engineer
- Project coordinator
- Design manager
- Field engineer
- Office engineer
- Quantity surveyor
- Project engineer
- Area superintendent
- Project superintendent
- Estimator
- Lead estimator
- Senior estimator
- Chief estimator
- Site Manager
- etc*

OBTAINING THE PROJECT

Bids

A bid is given to the owner by construction managers that are willing to complete their construction project. A bid tells the owner how much money they should expect to pay the construction management company in order for them to complete the project.

- Open bid: An open bid is used for public projects. Any and all contractors are allowed to submit their bid due to public advertising.
- Closed bid: A closed bid is used for private projects. A selection of contractors are sent an invitation for bid so only they can submit a bid for the specified project.

Selection methods

- Low-bid selection: This selection focuses on the price of a project. Multiple construction management companies submit a bid to the owner that is the lowest amount they are willing to do the job for. Then the owner usually chooses the company with the lowest bid to complete the job for them.
- Best-value selection: This selection focuses on both the price and qualifications of the contractors submitting bids. This means that the owner chooses the contractor with the best price and the best qualifications. The owner decides by using a request for proposal (RFP), which provides the owner with the contractor's exact form of scheduling and budgeting that the contractor expects to use for the project.
- Qualifications-based selection: This selection is used when the owner decides to choose the contractor only on the basis of their

qualifications. The owner then uses a request for qualifications (RFQ), which provides the owner with the contractor's experience, management plans, project organization, and budget and schedule performance. The owner may also ask for safety records and individual credentials of their members.

Payment contracts

- Lump sum: This is the most common type of contract. The construction manager and the owner agree on the overall cost of the construction project and the owner is responsible for paying that amount whether the construction project exceeds or falls below the agreed price of payment.
- Cost plus fee: This contract provides payment for the contractor including the total cost of the project as well as a fixed fee or percentage of the total cost. This contract is beneficial to the contractor since any additional costs will be paid for even though they were unexpected for the owner.
- Guaranteed maximum price: This contract is the same as the cost-plus-fee contract although there is a set price that the overall cost and fee do not go above.
- Unitprice: This contract is used when the cost cannot be determined ahead of time. The owner provides materials with a specific unit price to limit spending.

PROJECT STAGES

Design

The design stage contains a lot of steps: programming and feasibility, schematic design, design development, and contract documents. It is the responsibility of the design team to ensure that the design meets all building codes and regulations. It is during the design stage that the bidding process takes place.

- Programming and feasibility: The needs, goals, and objectives must be determined for the building. Decisions must be made on the building size, number of rooms, how the space will be used, and who will be using the space. This must all be considered to begin the actual designing of the building.
- Schematic design: Schematic designs are sketches used to identify spaces, shapes, and patterns. Materials, sizes, colors, and textures must be considered in the sketches.
- Design development (DD): This step requires research and investigation into what materials and equipment will be used as well as their cost.

- Contract documents (CDs): Contract documents are the final drawings and specifications of the construction project. They are used by contractors to determine their bid while builders use them for the construction process. Contract documents can also be called working drawings.

Pre-construction

The pre-construction stage begins when the owner gives a notice to proceed to the contractor that they have chosen through the bidding process. A notice to proceed is when the owner gives permission to the contractor to begin their work on the project. The first step is to assign the project team which includes the project manager (PM), contract administrator, superintendent, and field engineer.

- Project manager: The project manager is in charge of the project team.
- Contract administrator: The contract administrator assists the project manager as well as the superintendent with the details of the construction contract.
- Superintendent: It is the superintendent's job to make sure everything is on schedule including flow of materials, deliveries, and equipment. They are also in charge of coordinating on-site construction activities.
- Field engineer: A field engineer is considered an entry-level position and is responsible for paperwork.

During the pre-construction stage, a site investigation must take place. A site investigation takes place to discover if any steps need to be implemented on the job site. This is in order to get the site ready before the actual construction begins. This also includes any unforeseen conditions such as historical artifacts or environment problems. A soil test must be done to determine if the soil is in good condition to be built upon.

Procurement

The procurement stage is when Labour, materials and equipment needed to complete the project are purchased. This can be done by the general contractor if the company does all their own construction work. If the contractor does not do their own work, they obtain it through subcontractors. Subcontractors are contractors who specialize in one particular aspect of the construction work such as concrete, welding, glass, or carpentry. Subcontractors are hired the same way a general contractor would be, which is through the bidding process. Purchase orders are also part of the procurement stage.

- Purchase orders: A purchase order is used in various types of businesses. In this case, a purchase order is an agreement between a buyer and seller that the products purchased meet the required specifications for the agreed price.

Construction

The construction stage begins with a pre-construction meeting brought together by the superintendent. The pre-construction meeting is meant to make decisions dealing with work hours, material storage, quality control, and site access. The next step is to move everything onto the construction site and set it all up. At this stage, construction monitoring and supervision is of great importance to ensure that a project is completed on time and on budget, while meeting all relevant regulations and quality standards.

Contractor progress payment schedule

A Contractor progress payment schedule is a schedule of when (according to project milestones or specified dates) contractors and suppliers will be paid for the current progress of installed work.

Progress payments are partial payments for work completed during a portion, usually a month, during a construction period. Progress payments are made to general contractors, subcontractors, and suppliers as construction projects progress. Payments are typically made on a monthly basis but could be modified to meet certain milestones.

Progress payments are an important part of contract administration for the contractor. Proper preparation of the information necessary for payment processing can help the contractor financially complete the project.

Owner occupancy

Once the owner moves into the building, a warranty period begins. This is to ensure that all materials, equipment, and quality meet the expectations of the owner that are included within the contract.

ISSUES RESULTING FROM CONSTRUCTION

Noise control

A lot of ear plugs are worn on the job site

Environmental protections

- Storm water pollution: As a result of construction, the soil is displaced from its original location which can possibly cause environmental problems in the future. Run-off can occur during storms which can possibly transfer harmful pollutants through the soil to rivers, lakes, wetlands, and coastal waters.
- Endangered species: If endangered species have been found on the construction site, the site must be shut down for some time. The construction site must be shut down for as long as it takes for authorities to make a decision on the situation. Once the situation

has been assessed, the contractor makes the appropriate accommodations to not disturb the species.

- Vegetation: There may often be particular trees or other vegetation that must be protected on the job site. This may require fences or security tape to warn builders that they must not be harmed.
- Wetlands: The contractor must make accommodations so that erosion and water flow are not affected by construction. Any liquid spills must be maintained due to contaminants that may enter the wetland.
- Historical or cultural artifacts: Artifacts may include arrowheads, pottery shards, and bones. All work comes to a halt if any artifacts are found and will not resume until they can be properly examined and removed from the area.

CONSTRUCTION ACTIVITY DOCUMENTATION

Project meetings take place at scheduled intervals to discuss the progress on the construction site and any concerns or issues. The discussion and any decisions made at the meeting must be documented.

Diaries, logs, and daily field reports keep track of the daily activities on a job site each day.

- Diaries: Each member of the project team is expected to keep a project diary. The diary contains summaries of the day's events in the member's own words.
  They are used to keep track of any daily work activity, conversations, observations, or any other relevant information regarding the construction activities. Diaries can be referred to when disputes arise and a diary happens to contain information connected with the disagreement. Diaries that are handwritten can be used as evidence in court.
- Logs: Logs keep track of the regular activities on the job site such as phone logs, transmittal logs, delivery logs, and RFI (Request for Information) logs.
- Daily field reports: Daily field reports are a more formal way of recording information on the job site. They contain information that includes the day's activities, temperature and weather conditions, delivered equipment or materials, visitors on the site, and equipment used that day.

Labour statements are required on a daily basis. Also list of Labour, PERT CPM are needed for Labour planning to complete a project in time.

RESOLVING DISPUTES

- Mediation: Mediation uses a third party mediator to resolve any disputes. The mediator helps both disputing parties to come to a mutual agreement. This process ensures that no attorneys become involved in the dispute and is less time-consuming.

- Minitrial: A minitrial takes more time and money than a mediation. The minitrial takes place in an informal setting and involves some type of advisor or attorney that must be paid. The disputing parties may come to an agreement or the third party advisor may offer their advice. The agreement is Non-binding and can be broken.
- Arbitration: Arbitration is the most costly and time-consuming way to resolve a dispute. Each party is represented by an attorney while witnesses and evidence are presented. Once all information is provided on the issue, the arbitrator makes a ruling which provides the final decision. The arbitrator provides the final decision on what must be done and it is a binding agreement between each of the disputing parties.

TERMINOLOGY

The following terms are commonly used in the industry:

- Construction management (CM) refers to form of delivery and (in the UK) management of the site.
- Real estate management (REM) is professional property advice (a continuous process, as opposed to a process).
- Corporate real estate management (CREM) is a company-focused variant.
- Management contracting (MC):
- UK: form of delivery
- US: CM at risk
- Programme management (ProgM) is concerned with management of a client's portfolio (the programme, in this sense, is equivalent to a client's brief) or (in the UK) managing time in a project.
- Project control (PC) is the tracking and reporting of the progress, time, cost and quality of a project. This function can be characterized as passive, whereas construction project management (CPM) is active.
- Project leader (PL): The person responsible for achieving the project's objectives; acts as a manager "in-line".
- Project director (PD): The leader of a large project that can be broken down into sub-projects (*e.g.* the Channel tunnel) or the head of a PM organization
- Owner representative (OR): The representative of the owner; may be internal or external to the company
- Document Control (DC): Key function of a project manager
- Build operate transfer (BOT)
- Finance build operate transfer (FBOT)

- Design build operate transfer (DBOT)
- Build own operate (BOO)
- Engineering procurement construction (EPC)
- Private finance initiative (PFI)
- General contract (GC)
- Joint Venture (JV): A collaboration between two or more companies from the same or different backgrounds and/or fields to complete a common project. Joint Ventures typically have a lead contractor that deals with most of the business aspects. The lead will usually have a bigger stake in the partnership (*i.e.* the lead will have a 65per cent stake while the second contractor might have 30per cent and a third might have 5per cent).
- Guaranteed maximum price (GMP): A contract-specified upper limit to the project's final price
- Multiple prime contracts (MPC)
- UK: One contractor takes responsibility for the development (package deal).
- US: A client may have five or six prime contractors.

STUDY AND PRACTICE

Construction Management education comes in a variety of formats: formal degree Programmes (Two-year associate degree; four-year baccalaureate degree, masters degree,project management, operations management engineer degree, doctor of philosophy degree, postdoctoral researcher); on-the-job-training; and continuing education and professional development. Information on degree Programmes is available from the American Council for Construction Education (ACCE) or the Associated Schools of Construction (ASC).

According to the American Council for Construction Education (the academic accrediting body of construction management educational Programmes in the U.S.), the academic field of construction management encompasses a wide range of topics. These range from general management skills, through management skills specifically related to construction, to technical knowledge of construction methods and practices. There are many schools offering Construction Management Programmes, including some offering a Masters degree.

CONSTRUCTION MANAGEMENT SOFTWARE

Capital project management software (CPMS) refers to the systems that are currently available that help capital project owner/operators, Programme managers, and construction managers, control and manage the vast amount of information that capital construction projects create. A collection, or portfolio of projects only makes this a bigger challenge. These systems go by different names: capital project management software, construction management software, project management information systems. Usually Construction Management can be referred as subset of CPMS where the scope of CPMS is not limited to construction phases of project. There are systems available from vendors such as Aurigo Software which will not only manage capital projects but also the entire project portfolio (Programme).

There is a belief among many senior executives that by investing in an ERP system, their operational problems in project delivery will become history. That is hardly ever the case. ERP systems are an essential part of running a large and complex organization and they are excellent "systems of record" but are not sufficient tool for managing large capital projects and obtain real-time information and analytics around it. There is a growing trend of capital project owners realizing that ERP are not designed for the needs of capital project management and what is required is a CPMS system, which is integrated with ERP and other internal systems such as GIS, Mapping, etc. CPMS is a critical component of overall IT strategy for Capital project owners and capital project owners are realizing it now, more than ever.

BUSINESS MODEL

Fig. Skyscrapers under construction inPanama City, Panama

The construction industry typically includes three parties: an owner, a designer (architect or engineer) and a builder (usually known as ageneral contractor). There are traditionally two contracts between these parties as they work together to plan, design and construct the project. The first contract is the owner-designer contract, which involves planning, design and construction administration. The second contract is the owner-contractor contract, which involves construction.

An indirect third-party relationship exists between the designer and the contractor, due to these two contracts. An owner may also contract with a construction project management company as an advisor, creating a third contract relationship in the project. The construction manager's role is to provide construction advice to the designer, design advice to the constructor on the owner's behalf and other advice as necessary.

DESIGN, BID, BUILD CONTRACTS

The phrase "design, bid, build" describes the prevailing model of construction management, in which the general contractor is engaged through a tender process after designs have been completed by the architect or engineer.

DESIGN-BUILD CONTRACTS

Many owners – particularly government agencies – let out contracts known as *design-build contracts*. In this type of contract, the construction team (known as the design-builder) is responsible for taking the owner's concept and completing a detailed design before (following the owner's approval of the design) proceeding with construction. Virtual design and construction technology may be used by contractors to maintain a tight construction time.

There are three main advantages to a design-build contract. First, the construction team is motivated to work with the design team to develop a practical design. The team can find creative ways to reduce construction costs without reducing the function of the final product. The second major advantage involves the schedule.

Many projects are commissioned within a tight time frame. Under a traditional contract, construction cannot begin until after the design is finished and the project has been awarded to a bidder. In a design-build contract the contractor is established at the outset, and construction activities can proceed concurrently with the design. The third major advantage is that the design-build contractor has an incentive to keep the combined design and construction costs within the owner's budget.

The major problem with design-build contracts is an inherent conflict of interest. In a standard contract the designer is responsible to the owner to review the builder's work, ensuring that the products and methods meet specifications and codes.

An independent builder may pick up design flaws which might go unnoticed (or unmentioned) if the builder is also the designer. The owner may get a building that is over-designed to increase profits for the design-builder, or a building built with lesser-grade products to maximize profits. If speed is important, design and construction contracts can be awarded separately; bidding takes place on preliminary plans in a not-to-exceed contract instead of a single, firm design-build contract.

PLANNING AND SCHEDULING

Project-management methodology is as follows:

- Work breakdown structure
- Project network of activities
- Critical path method (CPM)
- Resource management
- Resource leveling
- Risk assessment

ARCHITECTURE–ENGINEER

- Work inspection
- Change orders
- Review payments
- Materials and samples
- Shop drawings
- Three-dimensional image

AGENCY CM

Construction cost management is a fee-based service in which the construction manager (CM) is responsible exclusively to the owner, acting in the owner's interests at every stage of the project. The construction manager offers impartial advice on matters such as:

- Optimum use of available funds
- Control of the scope of the work
- Project scheduling
- Optimum use of design and construction firms' skills and talents
- Avoidance of delays, changes and disputes
- Enhancing project design and construction quality
- Optimum flexibility in contracting and procurement
- Cash-flow management

Comprehensive management of every stage of the project, beginning with the original concept and project definition, yields the greatest benefit to owners. As time progresses beyond the pre-design phase, the CM's ability to effect cost savings diminishes. The agency CM can represent the owner by helping select the design and construction teams and managing the design (preventing scope creep), helping the owner stay within a predetermined budget with value engineering, cost-benefit analysis and best-value comparisons. The software-application field of construction collaboration technology has been developed to apply information technology to construction management.

CM AT-RISK

CM at-risk is a delivery method which entails a commitment by the construction manager to deliver the project within a Guaranteed Maximum Price (GMP). The construction manager acts as a consultant to the owner in the development and design phases (preconstruction services), and as a general contractor during construction. When a construction manager is bound to a GMP, the fundamental character of the relationship is changed. In addition to acting in the owner's interest, the construction manager must control construction costs to stay within the GMP.

CM at-risk is a global term referring to the business relationship of a construction contractor, owner and architect (or designer). Typically, a CM at-risk arrangement eliminates a "low-bid" construction project. A GMP agreement is a typical part of the CM-and-owner agreement (comparable to a "low-bid" contract), but with adjustments in responsibility for the CM. The advantage of a CM at-risk arrangement is budget management. Before a project's design is completed (six to eighteen months of coordination between designer and owner), the CM is involved with estimating the cost of constructing a project based on the goals of the designer and owner (design concept) and the project's scope. In balancing the costs, schedule, quality and scope of the project, the design may be modified instead of redesigned; if the owner decides to expand the project, adjustments can be made before pricing. To manage the budget before design is complete and construction crews mobilized, the CM conducts site management and purchases major items to efficiently manage time and cost.

ACCELERATED CONSTRUCTION TECHNIQUES

Starting with its Accelerated Bridge Programme in the late 2000s, the Massachusetts Department of Transportation began employing accelerated construction techniques, in which it signs contracts with incentives for early completion and penalties for late completion, and uses intense construction during longer periods of complete closure to shorten the overall project duration and reduce cost.

SAFETY MANAGEMENT DURING CONSTRUCTION

SCOPE OF SAFETY MANAGEMENT

Safety management during construction phase covers:

- Planning of work to avoid personal injury and property damage
- Monitoring of work to provide early detection and correction of unsafe practices and conditions
- Protecting adjacent public and private properties to provide for the safety of the public

- Providing safety education and incentive programmes
- Complying with federal and state Occupational Health and Safety Acts (OSHA).

Roles of the Agency in Safety Management

The Agency's role is to establish awareness that the prevention of accidents and protection of employees, the public, and property is a top priority. The Agency should have a safety management plan which can be a sub section of the PMP or on larger projects a separate subsidiary planning document. The requirements of the safety management plan should be incorporated as part of the contract documents.

The Agency through their CM monitors individual contractor safety performance for compliance with the above contractual safety requirements and conducts regular contractor safety audits and loss control surveys. When a violation of job safety is observed the CM will advise the Agency to notify the contractor in writing to correct the violation.

Roles of the Contractor in Safety Management

Contractors are responsible for having a safety management plan in place and for assuring safety on site, the safe and healthful performance of their work, preventing accidents or damage to adjacent public and private property, and safety training of their employees. When a contractor is advised by the Agency of a safety violation, the contractor should respond in writing and immediately take corrective action as set out in their safety management plan.

Enforcement

Contractors enforce safety by developing a Job Hazard Analysis for the work to be undertaken and discussing actions needed to provide safety at jobsite planning meetings. Supervisors draw on their safety experience to direct the actions of those under their direction. Contractor staff should include a safety professional who undertakes surveillance of operations to eliminate sources of potential accidents.

Education

Contractors give newly employed, promoted, and/or transferred personnel comprehensive safety indoctrination on topics such as: workplace hazards, required protective equipment, procedures for reporting unsafe job conditions, procedures for reporting accidents, contractor job rules, location of first-aid and medical facilities, and tool box safety meeting requirements. Safety should be a standing item at site meetings. Foremen or shift supervisors should also hold regular crew training (toolbox) meetings to cover specific safety procedures pertinent to the crew's on-going activity.

Incentives

Contractors should display signs and posters at the job site to reinforce safety training and as an incentive to maintain interest in job safety with the changing work assignments and jobsite conditions. The Agency should encourage contractors to introduce employee incentive programmes that reward safe work performance through personal recognition and prizes such as belt buckles, pins, or lunch boxes.

Audits

The Agency through the CM monitors and conducts regular audits of contractor safety performance and notifies the contractor in writing of unsafe practices observed. Should a contractor fail to correct an unsafe condition or practice, the CM may recommend that the Agency issue a stop work order until the condition is corrected.

Accident Investigation and Record Keeping

Accidents should be investigated without delay by the contractor and the investigation should generate recommendations for corrective actions to prevent recurrence of similar accidents. The contractor's accident report, project records, progress reports, and daily time reports may become important evidential material in any ensuing legal action. The contractor prepares monthly accident summary reports for submission to the CM. These reports will allow the CM to assess contractor safety performance as measured by recordable and lost time accident frequency rates and the type and cause of accidents. The federal and state regulations mandate reporting of certain injury accidents to the authorities.

SAFETY AND SECURITY CERTIFICATION

The word "safety" is used to deal with hazards (due to unintentional acts) and "security" to deal with vulnerabilities (due to intentional acts). The projects that are the subject of this Handbook fall mainly under bus transit mode. The standards that apply include: American Society of Heating, Refrigerating, and Air Conditioning Engineers (ASHRAE); Institute of Electrical and Electronics Engineers; and federal, city, and state building codes. In addition, for bus operations the project manager should consider regulations by the Federal Motor Carrier Safety Administration (FMCSA), Occupational Safety and Health Administration (OSHA), FTA Drug and Alcohol, and the applicable State's Department of Motor Vehicle regulations. The project manager, with support from the commissioning manager and led by the safety and security personnel, should consider self-certification of the project for satisfactory compliance with a formal list of hazards and applicable safety and security requirements. The self-certification will document all actions taken, identify and close gaps, and assure the stakeholders of the Agency's commitment to safety and security. The responsible parties self-certify by signing off on this document.

In addition, the Agency should establish strong ties with emergency response agencies and resources to provide mutual assistance in an event of major emergencies on or near the project. Emergency preparedness requires working with local emergency management groups to develop procedures and contingency plans specific to the location and the nature of the project, and perform specific drills to simulate emergencies.

OPERATIONAL AND MAINTENANCE MANUALS

The project manager must assure that the technical requirements for OandM manuals are addressed by the designers of record and specify schedule requirements for submittal of the OandM manuals well in advance of construction completion. The contractor prepares and submits the OandM manuals in accordance with specification requirements. The specifications should address the systems and subsystems for which manuals are required. It should outline the media (hard copy, CDROM, etc.), the quantities, formatting, and schedule for delivery. You should consider liquidated damages to be specified and assessed to the contractor if the OandM manuals and follow-on training are not provided within the specified time frame well in advance of the operations. The OandM manuals typically contain.

TRAINING AND TRANSITION TO OPERATIONS

Under general direction of the project manager and through supervision of the commissioning manager, training is conducted by the contractor or its subcontractors or suppliers and takes place after the acceptance of equipment and OandM manuals, and before operations begin. The project manager will work closely with the commissioning manager to schedule personnel in training classes. When a large number of personnel are involved, the project may choose to have several experienced personnel with ability to teach, take part in train-the-trainer sessions (if available) and begin training others. The contractor must submit a detailed outline of the training programme with the OandM manuals. The commissioning manager will review the material for compliance with construction specifications and get input from the specific group being trained. The commissioning manager with assistance from the contractor and the group being trained will ensure that all training requirements are met and the training material is turned over to the Agency.

AS-BUILT DOCUMENTATION

The project manager should assure the construction contract calls for the contractor to mark up the changes on the drawings and specifications as they occur during construction and turn in a set of marked up drawings and specifications.

Under general direction of the project manager, the construction manager verifies the construction contractor is keeping the drawings up to date. The project manager should provide in the agreement with the designer of record scope for the designer to review, approve, and produce the final as-built drawings.

The commissioning manager will use the as-built documentation as necessary to commission the project. As-built drawings will save significant life cycle costs and avoid potential safety hazards during operations and life of the facilities. The commissioning manager will review the final as-built drawings and assure they are a part of the final commissioning report.

WARRANTY ADMINISTRATION

Warranty is an assurance by the manufacturer in writing to the Agency that assumes stipulated responsibility for the performance of the equipment supplied for a specified period after acceptance. Warranty lengths typically vary in length from six months to five years.

The project manager should be careful not to accept a piece of equipment before it is ready for overall operations. The project manager should consider provisions for extended warranty when equipment must arrive and must be accepted well in advance of total operations. Under general direction of the project manager, the commissioning manager will work closely with the procurement department personnel who will check the equipment warranties against contractual requirements.

ROLE OF THE AGENCY IN PROJECT COMPLETION

Closing contractual activities requires the Agency's project manager to oversee final settlement of project contracts, acceptance of contract deliverables, collection of contract documents and records (such as as-built drawings, operation and maintenance manuals, and warranties, etc.), and approval of final payments. The project manager's responsibilities for administrative closeout relate to demobilizing the project team and completing activities with other stakeholders, arranging the disposition of project records, closing of funding and financing agreements, and performing an evaluation of project success and lessons learned.

CONTRACTUAL CLOSEOUT

Construction Contracts

The project manager, commissioning manager, construction manager/ resident engineer, and contract administrator, should follow the procedures and actions specified in each contract's terms and conditions to settle and close the project's construction contract agreements.

For a typical construction contract you will need to confirm the completion and acceptability of the following activities:

Manuals and Training – The contractor delivers the operations and maintenance (OandM) manuals for the facilities constructed and equipment installed and provides any associated training of Agency staff in their use.

Beneficial Occupancy – A contract is substantially complete when the permitting authority issues a Certificate of Beneficial Occupancy to the Agency and then the Agency can occupy and begin use of the facility and equipment. It is important on taking beneficial occupancy that you ensure the construction manager/resident engineer prepares a punch list of open items for the contractor to complete.

Guaranties and Warranties – With beneficial occupancy confirm that the contractor has initiated the guaranties and warranties associated with the facility and equipment.

Record or As-built Drawings – The construction manager/resident engineer confirms that the contractor has submitted the record drawings that show the as-built condition of the constructed facility and installed equipment.

Final Inspection – Lead a final walk through inspection of the facility to confirm that the contractor has completed the open punch list items and all work is completed correctly and satisfactorily.

Resolve Outstanding Change/Claim Disputes – You should make every effort to resolve any outstanding contract disputes so that they do not drag on past contract and project completion.

Final Payment – With the above activities satisfactorily completed you can approve the final payment to the contractor and the Agency can close the contract.

Commissioning – Assure that all other commissioning activities have been completed in a satisfactory manner.

Professional Service Contracts

Although closing a professional service contract, for such as design or management services, does not involve as many milestones and activities as a construction contract, you still need to follow the completion procedures dictated by the terms of the professional service contract.

Typical closeout activities for a professional service contract include:

Verification of Scope Completion – Confirm that the professional service contractor has satisfactorily delivered the services called for in the contract scope of work.

Contract Audit – Where contract payments are on a cost plus fee basis, the contract provisions should give the Agency the right to audit the contractor's costs. The audit should verify items such as direct labour rates, support for time charges, support for other direct costs, and justification for overhead rates.

Final Payment and Release of Retention – With scope completed satisfactorily and audit completed, you can approve the final payment and release of any retention held back from prior contract payments pending satisfactory completion of services and audit of costs.

ADMINISTRATIVE CLOSEOUT

Project Demobilisation

Managing the demobilisation of the project can often test the project manager's administrative and interpersonal skills in order to address the following end of project issues:

- Key project staff that see the end of the project coming and acquire positions elsewhere in the Agency before their project role and duties are complete.
- Or the opposite, Agency project staff whose duties are complete and are difficult to relocate off the project because there is no immediate position for them to move to.
- Professional service consultants, whose role is concluding, prematurely transfer key staff to newer long term assignments and/ or endeavor to stretch out their services to maintain revenue.

To manage demobilisation as project manager, you should develop a staffing plan for the final phase of the project that plans the reduction in the Agency's own forces and those of the professional service consultants. You should work with the Agency's human resources to help manage the transition of staff off the project, ease their anxieties surrounding what their next role is, and provide incentives for key staff to remain on the project and defer moving on. Similarly you should meet with the principals of the professional service firms to reach agreement on a timetable for winding down their services as the project concludes.

The project manager's final challenge once the demobilisation plans are in place is to keep the project team's attention focused on the tasks needed to complete the project as opposed to what they will be doing once the project is over.

Closure of Project Financing and Funding

The project manager will need to work with the Agency's finance staff to closeout the funding to the project. Where a project receives FTA funds, Circular 5010.1C Grant Management Guidelines sets out the activities required of an Agency to satisfy the FTA that all the Agency's project responsibilities and work are completed and the associated financial records closed.

Disposition of Project Records

The project manager has to arrange for project records to be transferred to the Agency's document control function. Project records required to be maintained will be determined by a combination of the Agency's own records retention policy, retention requirements imposed by parties funding the project such as the FTA, and any special requirements due to contract provisions. Should there be an unresolved change/claim dispute, it is important that all records pertaining to the contract and dispute also be retained.

Project Evaluation

Before the project is over and key project staff has dispersed, it is desirable for the project manager to hold a lessons learned session. The lessons learned should focus on identifying project strengths and weaknesses with recommendations on how to improve future performance on projects. To be most effective the project manager should encourage the project team to identify and document lessons learned through the project life cycle so that a database of lessons learned experience is built up for consideration at the final lessons learned session. Lessons learned can consider technical, managerial, and process aspects of the project.

5

Building Engineering

INTRODUCTION

Creating innovative and practical building engineering designs, our experts ensure that every project meets and exceeds the expectations of clients and building users.

Combining local market knowledge with technical expertise, our multidisciplinary approach to building design is informed by the client's needs and business case to produce cost-effective, functional, inspiring solutions.

We are known internationally for our creative, sustainable and holistic design approach which, from the outset, embraces low-carbon and economic operational performance. Our professionals around the world strive to delivery visionary buildings that maximize naturally occurring energy and minimize waste. From creation to completion and beyond, we help clients reach the optimum potential of every project.

Collaborating across a wide range of building design skills, and sharing knowledge and a passion for a sustainable built environment through our international network of offices, Building Engineering at AECOM creates designs that work for today and tomorrow.

The Civil Engineering profession is concerned with the built environment. Civil engineers plan, design, and construct major facilities, including highways, transit systems, airports, dams, water and wastewater treatment systems, tunnels, energy facilities, harbors, canals, buildings, and bridges. Civil engineers manage our air, water, and energy resources and protect society from natural catastrophes, such as earthquakes, and the hazards society itself generates in the form of toxic wastes.

Because these functions are often crucial to the day-to-day lives of most people and the facilities involved are physically substantial, civil engineers bear an important responsibility to the public. Their role is often more than just technical, requiring also a high degree of communicative skills and an ability to deal with people. Civil engineers can be found in industry, consulting firms, and government. This is one of the few areas of engineering in which the engineer often deals directly with the public and public agencies in every phase of major infrastructural projects.Our Civil and Environmental Engineering Department offers specialisation in environmental, geotechnical, structural, transportation, and water resources engineering.

Environmental Engineering involves sustainable design for the control and protection of the environment and its resources. Environmental engineers design systems for water quality and treatment, wastewater treatment, hazardous waste management and treatment and control of air pollutants.

Geotechnical Engineering encompasses the areas of soil mechanics and foundation engineering. It is concerned with design and construction of structures built on or below the ground surface and the physical characteristics of soil and rocks or composite material. Geotechnical engineering includes the design of foundations of bridges and buildings, design of tunnels and dams, and the geological factors affecting all structures.

Structural Engineering includes the design of all types of structures including buildings, dams, bridges and tunnels and the monitoring of their construction. A primary concern of structural engineers is predicting the loads that a structure will have to resist during its life and ensuring that it will be both safe and useful.

Transportation Engineering deals with the planning, design, construction and operation of highways, railways, air transportation systems, and their terminals. Transportation engineers are involved in the total transportation system, including the planning, design, implementation, administration, management, and performance evaluation. Highway engineering and traffic engineering are subfields of transportation engineering that involve highway design, traffic operations, and highway safety.

Water Resources Engineering involves flood control, harbour and river development and water quantity management. It also includes hydrology, which encompasses the occurrence, distribution, movement and properties of the waters of the Earth and their environmental relationships.

WHAT IS BUILDING ENGINEERING?

Building Engineers are concerned with the planning, design, construction, operation, renovation, and maintenance of buildings, as well as with their impacts on the surrounding environment. Building Engineering, commonly known in the US as Architectural Engineering, is an interdisciplinary programme that integrates pertinent knowledge from different disciplines:

- Civil engineering for building structures and foundation;
- Mechanical engineering for Heating, Ventilation and Air-Conditioning system (HVAC), and for mechanical service systems;
- Electrical engineering for power distribution, control, and electrical systems;

- Physics for building science, lighting and acoustics.
- Chemistry and biology for indoor air quality;
- Architecture for form, function, building codes and specifications;
- Economics for project planning and scheduling.

The building engineer explores all phases of the life cycle of a building and develops an appreciation of the building as an advanced technological system. Problems are identified and appropriate solutions found to improve the performance of the building in areas such as:

- Energy efficiency, passive solar engineering, lighting and acoustics;
- Indoor air quality;
- Construction management
- HVAC and control systems
- Advanced building materials, building envelope
- Earthquake resistance, wind effects on buildings, computer-aided design.

Why Choose Building Engineering?

While there are many universities that offer Building/Architectural Engineering programmes, Concordia's Building Engineering Programme remains unique in Canada. The programme is designed to meet the needs of the construction industry by providing engineers familiar with the overall design of built facilities. To reflect the increasing dependence of computer technology in engineering, the Department is now offering an Information Technology Option in the Building Engineering programme.

Students in the graduate programmes will have the opportunity to develop expertise in one or more of the following areas of specialisations:

- Building Indoor Environment covers the environmental aspects in the design, analysis and operation of energy-efficient, healthy and comfortable buildings. Fields of specialisation include: thermal comfort, air quality, lighting, acoustics, HVAC and control systems.
- Building Envelope is an application area which draws from all areas of building engineering, especially building science and indoor environment. It focuses on the analysis and design of building envelopes, including durability, heat and moisture transfer and interaction with the indoor environment.
- Building Science focuses on the analysis and control of the physical phenomena affecting the performance of building materials and building enclosure systems.
- Building Structure area concerns with the principles of structural mechanics, material Behaviour and their applications to the analysis and design of steel, reinforced concrete and timber building structures. Fields of specialisation include: wind and seismic effects on buildings.

- Construction Management includes construction techniques, construction processes, planning, scheduling; project tracking and control, labour and industrial relations, and legal issues in construction.
- Computer Aided Engineering is an exciting area in Engineering. Even though computers have become ubiquitous in the architecture-engineering-construction industry, their present use is mostly limited to drafting, analysis, member sizing, cost estimation, and construction management. Computers have tremendous untapped potentials in the field of Building Engineering.
- Energy Efficiency is also an application area which draws from the building science and building environment areas. It includes analysis, design, and control of energy-efficient buildings and HVAC systems, solar energyutilisation and intelligent buildings.

BUILDING STRUCTURES

Structural engineering includes all structural engineering related to the design of buildings. It is the branch of structural engineering that is close to architecture.

Structural building engineering is primarily driven by the creative manipulation of materials and forms and the underlying mathematical and scientific ideas to achieve an end which fulfills its functional requirements and is structurally safe when subjected to all the loads it could reasonably be expected to experience.

This is subtly different from architectural design, which is driven by the creative manipulation of materials and forms, mass, space, volume, texture and light to achieve an end which is aesthetic, functional and often artistic.

The architect is more than the lead designer on buildings, with a structural engineer employed as a sub-consultant.

The degree to which each discipline actually leads the design depends heavily on the type of structure. Many structures are structurally simple and led by architecture, such as multi-storey office buildings and housing, while other structures, such as tensile structures, shells and grid shells are heavily dependent on their form for their strength, and the engineer may have a more significant influence on the form, and hence much of the aesthetic, than the architect.

The structural design for a building must guarantee that the building is able to stand up safely, able to function without excessive deflections or movements which may cause fatigue of structural elements, cracking or failure of fixtures, fittings or partitions, or discomfort for occupants. It must account for movements and forces due to temperature, creep, cracking and imposed loads. It must also ensure that the design is practically buildable within acceptable manufacturing tolerances of the materials. It must allow the architecture to work, and the building services to fit within the building and function that includes:

- air conditioning
- ventilation
- smoke extract
- electric
- lighting

The structural design of a modern building can be extremely complex, and often requires a large team to complete.

Safety Assessment in the Built Environment

Safety on the jobsite is an important aspect of the overall safety in construction. Construction sites are in a constant state of change, dictating frequent inspections. The construction field is one of the most hazardous industrial fields.

The practice of safety in construction in the United States is regulated by governmental agencies such as the Occupational Safety and Health Administration (OSHA), which provides strict rules and regulations to enforce safety and health standards on jobsites. However, the practice of safety in Saudi Arabia is not regulated by any government agency as is the case in the United States. The prevention of accidents is, therefore, an area of responsibility of the top management of the organisation. Some construction companies realise the importance of reducing their accident rates not only for humanitarian reasons, but also because of the many financial benefits which flow from the safe conduct of the work.

Other companies do not have a strong belief in safety. This has serious repercussions when any unfortunate incidents occur. Good management should always insist that every engineer, supervisor, and laborer must be familiar with all basic safety aspects and practices that guard those around the sites from accidents and injuries.

The responsibility for safety on any construction project should be shared between all the parties involved in the project, namely, the owner, the designer or the architect, and the contractor. The owner as part of his safety responsibilities must ensure that the designer designs a safe project. He must also ensure that the contractor has a safety programme.

The owner should include the safety programme as an element of the bidding technicalities. The architect or the designer contributes towards ensuring the safety of the project by properly designing the temporary work and the permanent work from the safety point of view. The temporary works must be designed so that they provide a safe means of access to and about the construction work. The permanent work must be designed so that it is stable and safe for the users. Contractors should provide a safe environment for workers by meeting all safety requirements during construction processes, beginning with site preparation and ending with completion of the work. The safety performance of the company should be measured so that unsafe conditions and unsafe practices can be identified. There are several methods of measuring the safety performance on a job site. One way is by conducting a safety audit. A comprehensive audit is basically a revision of all aspects of the company's safety programme.

A properly conducted safety audit will determine the strengths and weaknesses of the current safety programme. Any problem areas that might adversely affect the success of the programme will be identified. Another way of measuring safety performance is by applying the concept of profiling which consists of the development of a corporate safety performance standard in a number of categories that are considered to be important.

Companies are then compared to this standard and a profile is made showing this comparison. The injury frequency rate which is the number of lost-time injuries per million man hours of exposure can also be used to measure safety performance. Jannadi and Al-Sudairi (1995) measured safety performance in Saudi Arabia using the injury frequency rate, and concluded that safety performance is best in the larger construction firms. Safety performance can also be measured by using a standard checklist that covers all aspects of the work.

Safety management is an approach aiming to remove or minimize the forces which cause losses by injuring the workers, or by damaging equipment and facilities. There are two models which study the causation of accidents. These are the behavioral and the situational models. The behavioral models consider humans as the major factor, responsible for accidents, whereas situational models consider the interactions between humans, environment, and the situation for studying the accident process. The causes of accidents are divided into two categories, immediate causes and contributing causes. The contributing causes of accidents include the physical condition of the workers and the management policies.

The immediate causes of accidents include unsafe acts and unsafe conditions. Unsafe conditions are physical conditions, which if left uncorrected, are likely to cause an accident.

Therefore, to improve safety on the work site, such conditions must be detected before an accident occurs. Unsafe conditions are detected on a work site by performing periodic safety inspections. These are usually conducted by the site safety personnel. Check lists are used to conduct such inspections. One such checklist is used to conduct this survey. This checklist tries to assess the safety of the site by considering only the unsafe conditions on the work site.

OBJECTIVE OF THE PAPER

The objective of this paper is to assess the level of safety practiced at different construction projects in Saudi Arabia. A survey of different construction sites was conducted, to determine whether safety levels differ according to the size of the project. A standard checklist was used to assess the safety practices of the different firms.

Research Methodology

A survey of 14 randomly selected projects was conducted to assess safety practices in Saudi Arabia. The sites were located in the Eastern province of Saudi Arabia. A standard checklist developed by the Saudi Arabian Oil Company (Saudi Aramco) for evaluation of safety was used in this survey.

The projects for the survey were selected at random, based on the fact that they were under construction at the time of the survey. The survey included two types of projects: i) large construction projects with estimated construction costs of over SR 50 millions (1 US $ = 3.75 SR); and ii) small construction projects with estimated construction costs of less than SR 5 million. There were 7 large projects and 7 small projects in all. The large projects included several office and commercial buildings. The small construction sites consisted mainly of residential buildings and housing.

The checklist used for the survey. This checklist includes those items which are perceived to be important from the safety point of view on the construction site. The checklist consists of 18 divisions and 96 items distributed among the different divisions. These are fire prevention, scaffold/mobile tower, cartridge operated tools, trenching excavation, housekeeping, sandblasting, power tool machine and equipment, heavy equipment, gas/electric welding, construction formwork, health and welfare, transportation, cranes and lifting devices, compressed gas, air compressors, site safety administration, temporary electric, and special items.

This checklist was developed by the Saudi Aramco. Saudi Aramco is an oil company considered to be one of the biggest in the world, with oil production averaging 8 million barrels per day. Saudi Aramco is a safety oriented oil company that applies both quantitative and qualitative risk evaluation techniques to its operating facilities.

Safety is monitored by conducting various safety inspections and reviews. In 1994, the numbers of both the planned quarterly safety inspections and the unscheduled inspections of the operating facilities and construction sites were 4,897 and 13,940 respectively. Also safety is actively promoted within Saudi Aramco through a wide range of publications, films and other media. Over 2.2 million safety awareness items were produced in 1994. The above checklist is used by Saudi Aramco to assess the safety on the construction projects undertaken by them.

The data was collected by a student of the authors. All sites were personally visited by the surveyor and the evaluations were made with the help of the project superintendent at the respective sites.

Table . Safety checklist

Divisions	Yes / No	Divisions	Yes / No
(1) Fire Prevention		**(2) Housekeeping**	
Adequate Fire Extinguisher		Site Access Roads	
Proper type Extinguisher		Security Fences/Gates	
Adequate Water Barrels /Buckets		Site Access Signs	
Properly located	*	Trash Container	
Emergency tel. # posted		Daily Clean-up	
Fire Watches		Materials Stacking	
Open Flame Operations		Aisleways	
Storage of Flammable /combustibles		Old Timber Denailed	
		Overall Condition	
Average Score		*Average Score*	
(3) Scaffold/Mobile Tower		**(4) Sandblasting**	
Planking		Operator's Hood (Air Supply)	
Wheel Locks		Air Filter (to hood)	
Using Proper Wheel/Condition		Hoses properly grounded	
Scaffold Access		Operator's protective clothing	
Proper Couplers		Remote area/warning signs	
Good ties/Outriggers			
Plumb and Level			
Condition of frame members			
Average Score		*Average Score*	
(5) Cartridge Operated Tools		**(6) Power Tools / Machine and Equipment**	
Penetration to safe zone		Properly guarded	
Control/Storage of cartridges		Tool Rest	
Proper maintenance of tools		Operator's Protective Equipment	
Certified operator		Damaged hand tools	
Operator Protective Equipment			
Average Score		*Average Score*	
(7) Excavation		**(8) Heavy Equipment**	
Shoving /Trench Box		Roll Over Protection	
Sloping		Back-up Alarms	
Barriers/Warning Signs/Lights		Licensed Operators	
Access/Egress			
Average Score		*Average Score*	
(9) Concrete Formwork		**(10) Gas/Electric Welding**	
Timber Adequate Strength		Proper Acetylene Press	
Protective Clothing & Equipt.		Acetylene on/off Wrench	
Firm Footing for Support		Operator's Protective Equip.	
Side Slope Bracing			
Average Score		*Average Score*	
Divisions	Yes / No	Divisions	Yes / No
(11) Health and Welfare		**(12) Compressed Gas**	

Limitations

Safety on the construction sites is divided into two aspects. They are the physical aspects and the behavioral aspect. The physical aspects of safety on the construction sites covers the unsafe work conditions. The behavioral aspects of safety cover the unsafe acts or behaviour on the part of the workers on the site. This paper concentrates only on the physical conditions on the site. The behavioral aspects are not covered. Hence the checklist used for this survey covers only those dimensions of safety which come under the domain of physical conditions.

Data Analysis

Each item within a division was evaluated as "yes" or "no" depending on its existence in the jobsite. Each "yes" was given a score of 100 and each "no" was given a score of zero. The division score was calculated by the following equation:

$$\frac{\sum[\#of"Yes"*100+\#of"No"*0]}{[\#ofapplicableitems]}$$

The division score was calculated by multiplying the number of "yes" by 100 and the total sum divided by the number of applicable items. Items not applicable for a particular project were ignored and not used in the calculation. Each project was scored by obtaining the average of the applicable division scores within that project. Projects were assessed based on the following scale:

0per cent-59per cent as poor

60per cent-69per cent as fair

70per cent-79per cent as good

80per cent-89per cent as very good

90per cent-100per cent as excellent.

Also the average of division scores was calculated for both small and large projects and ranked according to their safety score.

Test of Hypothesis

A critical value of rs is needed to test the alternative hypothesis that large and small projects generally agree on the important ranking of the divisions against the null hypothesis which says that there is no association between the rank. Using the table of critical value of Spearman's rank correlation coefficient, the critical value of rs with a = 0.05 and N = 16 is 0.425.

Since the calculated value of rs is larger than the critical value, then the null hypothesis will be rejected at the a = 0.05 level of significance. It appears that there is some agreement between the two ranks in both small and large projects.

GREEN ROOF DESIGN

Green roofs have been defined as water-proof structures that cover buildings with non-traditional means installed to improve environmental effects and better aesthetics. Green roofs have evolved from this definition into a whole new category of roofs that include anything from a roof with a small number of plants to a roof overlaid with a complete soil structure.

ADVANTAGES AND DISADVANTAGES OF GREEN ROOFS

Green roofs have a number of advantages. Most obviously, green roofs minimize the large amount of rain water Run-off that is a major energy concern in urban areas. Aesthetically, people using buildings with green roofs to enjoy having the option of sitting by the green roof overlook. Also, in some situations, patrons may be able to go onto the green roof as an alternative to sitting indoors. The minimisation of heat loss through the roof due to the extra insulation provided directly by the green roof is also important.

Table . Advantages of Green Roof Installations

Major Green Roof Advantages
1.) Reduce storm water run-off which in turn, reduces the stress on urban sewer systems and decreases run-off related pollution of natural waterways
2.) Insulating qualities mean reduced energy costs for building owners
3.) Air quality improvement – lower rooftop temperatures mean less smog from the "urban heat island effect."
4.) Noise pollution reduction –noise levels in a building can be reduced by as much as 40 decibels
5.) Extended life of the roof system due to moderated temperature swings that cause a roof system to expand and contract as well as protection from everyday wear and tear.

Despite the many benefits of installing a green roof, there can also be disadvantages. Roof leaks can be more common and can cause more damage. Maintenance can be a much more involved process than with a conventional roof. In addition to maintenance cost, the installation cost can become quite high. Also, the extra load needed to support the excess weight of a green roof may drive up the need for extra construction materials and cost.

Types of Green Roofs

There are two types of green roofs. The first is an intensive green roof. This type of roof must be designed for a larger load. Shrubs, small trees, and other plantings are all including in the design of an intensive green roof. More soil is used for this type due to more varied possibilities for plantings. The intensive green roof will have access provided for occupants of the building to enjoy the roof.

The other type of green roof is an extensive green roof. This type of roof requires less of a design load. The soil is much shallower, and the options for different plantings are limited. Access to the extensive roof should be restricted to necessary maintenance only. These are less visually attractive than intensive green roofs, but less energy is needed for irrigation purposes. Lastly, on some existing buildings, an extensive green roof will be the only option due to Pre-existing design load constraints.

Green Roof Materials

The materials of a green roof must be understood to implement feasibility of designs. It is typical for both types of roofs to have four main layers. The base layer is called the waterproofing layer. This layer is usually composed of a rubber or thermoplastic material. For a green roof, it is very important that this layer can hold up against the leachate from fertilizers or acid rain. Drainage layer should be provided for the roofs with slope angle less than 10°. Roofs with steeper angles can drain naturally due to gravity. This layer must capture and store both rain water and excess water from when the above plants are watered. It is very important that this layer not only holds water, but also allows it to drain so that the above soil does not expand.

The two layers above the drainage layer are design specifically to allow growth above. First, the root barrier membrane layer lies above the drainage layer. The root barrier layer provides protection to the waterproofing system by preventing the root migration into it. The root barrier consists of materials with the properties of elongation ability, tensile strength, resistance to cracking, resistance to puncture. The products used widely as root barriers in green roofs are RBM 400 and Blackline 500. The former consists of two layers of polyethylene reinforced with a multifilament grid. The later consists of two layers of pure polyethylene reinforced with low density polyethylene copolymer core and is of thickness 0.50mm.

The top layer is the soil or growing medium layer or the substrate layer. This layer holds specialised soil for plant growth. It is the component of the green roof laid on top of the root barrier layer. The basic requirements of the growing medium include having aged compost, no fertilizers required, a balanced pH, and bacteria which help with plant growth. The permeability range of good growing medium should be between 0.5 inches – 2 inches per hour. The purpose of the growing medium is not only to help the growth of the vegetation but also to support the basic ideas of green roof like retaining rain water and air, resist rot, provide good nutrients to plants during extreme cold and hot conditions. It provides stability to plants in windy areas. It must also not drastically increase the total weight on roof. The growing medium should consist of recycled materials and high quality compost so that it can act as self nutrient to plants. This reduces the maintenance cost of green roof. Natural soil cannot be used as the growing medium as they are heavy and cannot with stand weed growth. The different types of growing medium are: Lassenite Rotary Kiln Fired Pozzolan, Course sand, Reed/Sedge peat moss derived from natural resources. The substrate must consist of approximately 75per cent-80per cent of inorganic to 20per cent-25per cent organic compost. Higher organic content can lead to weed growth which is not advisable.

REASONING OF GREEN ROOFS INSTALLATIONS

Various studies have shown that waste treatment, water supply and regulation, erosion control, sediment retention, nutrient cycling, and climate regulation have all been very negatively affected by urbanisation. Since a good portion of these major environmental problems can be linked back to the transformation of natural landscapes to man-made impervious surfaces, it only seems natural that green roofs would be a popular way of turning some of these surface back to natural landscapes. The higher material costs; however, have slowed the spread of green roofs from entering mainstream urban construction.

Traditionally, a higher cost for green roofs was reported by taking only installation, maintenance, roof life, fees for storm water, and energy costs into account for the study of a single building.

Insulation Effect of Green Roofs

Even before this higher cost was reported, work had been done to determine energy savings within buildings that have green roofs installed. Thermal conductance in roofs is the key to energy savings. A huge proportion of heating or cooling energy escapes through the roof. When the thermal conductance of a roof is greater, more energy will escape and be wasted. Case studies were examined to give theoretical results for the thermal conductance and associated energy savings for buildings with green roof. Three different scenarios were examined.

It was found that green roofs can provide excellent insulation for non-insulated roofs. In fact the scenarios ranged from a 37per cent to a 48per cent total savings in energy when green roofs were installed on existing, non-insulated roofs. However, the energy savings were substantially less for moderately insulated roofs (4per cent-7per cent) and well insulated roofs (2per cent). In all cases; however, the green roof did provide energy savings. This large range of numbers for energy savings percentage opens up doubts about the higher long-term cost for green roofs that was reported. These doubts stem from uncertainties because it was only developed for a single building and the roof insulation issue was not addressed.

Air Quality Effect of Green Roofs

Green Roofs have been shown to cut down on air pollution in urban areas as well. This is another instance in which the higher green roof cost reported must be questioned. It did not take air pollution into account. Currently, reducing air pollution does not always lead to direct monetary costs. However, over thirty-seven hundred premature deaths in the United States can be directly related to increased air pollution levels of a single air pollutant, ozone. This country is increasingly becoming aware of this issue, and it should be considered that regulations in the near future may raise costs for building that do not take this into account.

An intensive study in Chicago analysed approximately 71per cent of the green roofs by area within the city. Ozone, nitrogen oxide, sulfur dioxide, and PM10 (particulate matter) were studied due to their high levels within the city of Chicago. The excessive plant growth alone removed 85 kilograms per year of the studied pollutants for each hectare installed of green roofs. This is very substantial. In fact, if the remaining traditional roof space in Chicago was converted to green roofing, around 2050 metric tons of pollutants would be saved from entering the air.

Installation Cost of Green Roofs

Despite energy savings and air pollution filtration benefits, the most important issue in the decision process often comes down to installation costs. Green roofing costs are approximately $150 per square meter of roofing to install. This exceeds the approximate $85 per square meter conventional roofs to be installed. However, green roofs can reduce operating costs. Since green roofs can cut down on energy costs by thermal conductance in a much more cost effective manor than insulation, insulation may not need to be installed. The construction savings from a lack of insulation begins to bring these per square meter installation costs closer for green and traditional roofs. When long term effects are thrown into the mix, green roofs can double roof life, reduce storm sewer discharge and size of pipe, and reduce energy. Taking all of these into account, green roof seem to be more economical in the long-term. Also, it was determined that social benefits can erase any life cycle cost difference between traditional and green roofs.

Environmental Benefit of Green Roofs

A green roof performs many functions that lead to an improvement of environmental performances of that building. These functions include absorption of rainfall, reduction of roof temperatures, improvement in ambient air quality, and provision of urban habitat. Timothy Carter conducted a benefit cost analysis (BCA) for the life cycle of an extensive roof and the results were compared with traditional roofing and the benefits and incentives for Run-off quantity and quality were determined.

Carter found that green roofs can function as a way to capture some of this rainfall into the soil and minimize the amount of Run-off produced. Carter investigated the feasibility of replacing flat roofs with a green roof system. The test roof was conducted on the campus of the University of Georgia. The roof was built so that it would be easily replicated. The roof used a root protection sheet, moisture retention mats, drainage panels, and geo-textile filter sheets. The greening on the roof all occurred within the first year. This analysis was estimated to take about 40 years and there would be a reroofing at 20 years. The benefits that were discovered from using a green roof included the following; "the roof life was doubled, the storm sewer pipe size was reduced, there was a reduction in the need for alternative storm water BMPs, the storm water utility fee was reduced, there was an energy savings, the habitat for insects and birds was increased, and there was a reduction in ambient air temperatures".

Lisa Kosareo also conducted a life cycle environmental assessment for green roofs. This assessment was done for intensive and extensive green roofs. Kosareo investigated a green roof project in Pittsburgh that compared the use of green roofs to a conventional roof. A life cycle analysis was conducted to compare the environmental aspects and potential impacts associated with constructing, maintain and disposing a green roof. The purpose of the LCA was to determine the option with the lowest negative impact. The LCA found that green roofs are the environmentally preferable choice when constructing a building, even with the need for extra resources. This is due to the amount of savings in energy and the increase life of the roofing membrane.

EXTENSIVE AND INTENSIVE DESIGN

The recent demand for more ecologically-minded urbanisation can be answered through the use of both intensive and extensive green roofs. Both employ the natural abilities of native plants to react as desired within the climate of the installation area. Also, soil layers are used to collect rainwater that would usually be wasted. The ecological difference between these two types of green roofs lies within the soil and the plants that are used.

Often, extensive green roofs use Sedum species, Phedimus or Hylotelephium for their majority of the vegetation. Intensive green roofs usually use grasses, perennial herbs and shrubs to make up the majority of their vegetation.

Extensive green roofs have substrate layers approximately 3–5 centimeters thick and a water-saturated weight of an approximate 50 kg/m2. Current substrates that are being used in Sweden have been based on natural soil mixes and improved with scoria or lava.

Substrates for green roofs are designed after taking into account system weight requirements, substrate water-holding capacity and oxygen diffusion to plant roots. Most plants used for extensive roofs are succulents that are able to store water in leaves or stems. This enables these plants to survive during the dry environments.

EFFECT OF GREEN ROOF ON RUN-OFF AND WATER QUANTITY

Many scientists have studied the correlation between precipitation, roof properties and Run-off. Jeroen Mentens obtained data on annual and seasonal Run-off from a literature that he performed. This data was collected from rainstorm events. A rainstorm is identified as a rainfall of 300 litre per second per hectare during 15 minutes.

The results from the literature review show that Run-off is determined by the roof type and may be as high as 91per cent for a traditional non-greened roof and as low as 15per cent for an intensive green roof. Therefore, green roofs can be used as tools to reduce the Run-off and ways to increase the retention time include green areas where water can infiltrate and evaporate. This is a tool that will use up unused space and therefore does not limit the demands of the people for open space on the ground.

Introduction to LEED Building Rating System

LEED (Leadership in Energy and Environmental Design) was developed as a national standard for building environmental minded and sustainable buildings. LEED was developed and continues to be administered by the U.S. Green Building Council. The first pilot version of LEED was released in 1999.

Sustainable sites, water efficiency, energy and atmosphere, materials and resources, indoor environmental quality, and innovation and design processes are the six major categories that the council concentrates upon when awarding certification. Overall, throughout all of these categories, points can be attained. The system is based around a total of 69 points.

There are four levels of certification for a building. LEED Certified is the first level. This level is attained by accumulating 26 points. LEED Silver may be achieved with 33 points. With 39 points, the third level, LEED Gold, is achieved. LEED Platinum is met with 52 points. A breakdown for the sectional points within the LEED system.

Table. LEED Sections and Points per Section

LEED-NC	Section Points
Sustainable Sites (SS)	14
Water Efficiency (WE)	5
Energy and Atmosphere (EA)	17
Materials and Resources (MR)	13
Indoor Environmental Quality (EQ)	15
Innovation and Design Process	5

A Local Case Study

Fig. Green Roof on Toledo-Lucas County Public Library consisting of Shrubs.

A case study of the public library located in downtown Toledo Ohio, which currently has a green roof installed, has been considered in order to gain an elaborated knowledge about the green roofs in the region of the proposed site. The Green Roof located at the library was planned for construction in 1992 and was completed and opened to public from 1993. It is the only library in Lucas

County consisting of a Green Roof. This green roof has been very aesthetically pleasing for both the workers and visitors to the library. Some of the shrubs and trees that are installed.

The details on the layout and construction of this green roof are especially interesting. The area of the roof library is 2 acres, of which 0.5 acres are covered with an intensive green roof.

As the green roof provided on the library was a part of the renovation programme, the roof was modified into grids. This grid system was designed in such a way that the system combines all the advantages of a loose-laid membrane installation with the added security of adhered membrane grid strips.

The grid strips compartmentalise the waterproofing system in to smaller areas, effectively limiting the scope of vegetated cover removal if a problem develops. Optional control drains are installed in each grid area as an active monitoring and altering mechanism. The drain opening is being used as an injection port to facilitate repair without vegetated cover removal. The Green Roof mostly consists of turf.

There has been some shrubs and trees installed to compliment the turf. Some of these shrubs and trees that have been installed. Additionally, the unique design of the roof which includes portions installed with a green roof and portions installed with access material.

Another unique design of this green roof. This green roof includes trees in specially designed planting areas allowing for additional soil coverage that is not included over the entire roof.

Fig. Green Roof on Toledo-Lucas County Public Library

A major construction material consideration is the SARNAFIL membrane. SARNAFIL is leading manufacturer of the waterproofing systems and high quality thermoplastic membranes. This membrane is made of thermoplastic PVC-polymer material. It is generally used in adhered systems. These membranes have a long life span as they are reinforced with fiberglass. The usage of fibre glass for reinforcement fortifies and offers exceptional dimensional stability and low coefficient of thermal expansion. The vapor

Fig. Trees grown in big troughs on the roof of Library.

permeability nature of this membrane avoids the water from the top soil to penetrate through the roof. The top layer of the membrane is coated with lacquer which gives the roof a self cleaning feature that resists staining from airborne dirt and pollutants. The other properties of this membrane are high flexibility, high mechanical strength, decay resistant, root resistant, weather and UV resistant.

Another important aspect of the design of this green roof is the sprinkler system that is fed by a potable water line. Due to the high interest in aesthetics that was incorporated into the roof design, the roof is watered in times of drought. This is not the most sustainable practice, but for this application it was necessary. In addition, the majority of the roof was designed with a drainage system. This minimizes flooding due to storm events. Although, this is a waste of the rainwater resource, this is necessary to allow continued access to the roof.

OBJECTIVES

The College of Engineering at the University of Toledo has interests in developing the North Engineering Building for office space, laboratories, and common use. The final goal is to bring all classes, professors, and students from the Scott Park Campus to the Main Campus. The architectural design firm SSOE has been hired to fast track the design plans for the North Engineering Building. Strict limits have been set on both the time restraints and budget of the project. The university has expressed distinct interests that all development, when possible, shall strive to attain the silver standards of the LEED certification.

The overall objective of this report was to present an engineering analysis for installing a green roof on the North Engineering Building. This analysis included far-reaching background information on green roofs. Also included is a recommendation on the layout of the green roof, the type of green roof, plants to be used, and other design details. To meet the objective of this report, an estimate of energy savings from green roof insulation and solar refraction, a comparative analysis of the life cycle cost of conventional and green roofs, and an estimate of the improvement of environmental quality have all been produced. The final aspect of the objective was to analyse LEED credits that can be attained for the North Engineering Building by installing a green roof.

ENGINEERING ANALYSIS

OVERVIEW OF ISSUES TO BE CONSIDERED

This project concentrated on the saw-toothed skylight region of the roof. Access to this portion could be provided from the upper catwalk that SSOE may plan to leave as a laboratory overlook. Since this area is not a flat roof, special considerations regarding slope, water collection, and available space have been explored.

Currently, the rainwater that hits the roof is sent into roof drains. Instead of collecting this water to be used for any variety of purposes, this water is wasted. To determine the affects of this waste, the volume of rainwater has been estimated. The roof area that has been looked at has a length of 240 feet and a width of 140 feet. The total area of this saw-toothed portion of the roof is a small amount more than 3730 square yards.

In addition to this area, there is a smaller flat area that will be visible and easily accessible from the catwalk area that SSOE is designing for. This area is at the same level along the southern portion of the western edge of the saw-toothed roof. At 80 feet by 160 feet, this area adds more than 1420 square yards for a total of more than 5150 total square yards of roof to catch Run-off from.

The ODNR report on average annual rainfall across Ohio has been used to find the rainfall in Toledo to be between 31-33 inches per year on average. This means that on average, each year sees over 4580 cubic yards of water wasted from the roof of the North Engineering Building area that was targeted for green roof installation. Currently, this water is channeled through the roof drain and ends up in either a sanitary or a storm sewer. This water was either discharged into a river or treated and then discharged. Either way, this was a waste of a source should be used more efficiently.

It was also important to be sure that the roof can hold a high intensity storm. If the original engineers designed the roof correctly, the existing roof drain system should have the capacity to unload the water from a high-intensity, low-recurrence storm. However, this storm was very important to analyse for the consideration of the installation of a green roof. It was found that, for a storm in Toledo with a recurrence interval of fifty years, 2.37 inches of rain falls in one hour or 5.05 inches falls over 24 hours. These two storms were needed to be the basis for design of the green roof water storage capacity.

A current consideration was the fact that the roof is designed with saw-tooth shaped windows. The windows face north. Although it seemed like sunlight could enter the building better if they were facing south, this leaves possible development on the southward facing side of the saw-teeth. Currently, the sunlight hitting this portion of the saw-teeth is wasted. The possibility of capturing this sunlight cannot be ignored when designing the roof for sustainable purposes. This report did not investigate the possibility of using sunlight as a source of alternative fuel. However, due to this possibility and the high slope of the saw-teeth shaped structures, no planting was to be installed on these slopes. This water was still be considered for collection, though.

Currently, it was determined that both rainwater and sunlight, which are naturally occurring, are being wasted by the set-up of this roof. This report addressed the benefits of installing a green roof, in addition to any other issues that arise during the project, to take advantage of maximizing the possible natural resources benefits.

Layout and Drainage Patterns

As indicated, no saw-toothed structures shall have the green roof materials installed on them. Instead, it was planned that all Run-off shall be directed to the areas between the saw-toothed structures where green roof materials shall be installed. The intentions of this was to allow all of the run-off the in the saw-tooth area of the roof to be collected by the green roof.

The slope of the roof shall not need to be changed for installation of the green roof. None of the existing roof drains shall be used under normal conditions. However, they shall remain intact. The case study examined in this report used a type of drainage system under the roof that would drain any additional water that cannot be stored in the planting or the soil. In the interest of wasting the least amount of rainwater possible, this technique is not be recommended. Instead, the existing roof drains that are to be left intact shall have their inlets raised to act as an emergency drain.

Rain gardens commonly use emergency drains as a means of capturing the maximum possible amount of rainwater without allowing excessive flooding. This design is similar to the design of raising the inlets of the existing roof drains. The fifty-year, twenty-four hour storm brings 5.05 inches of water. Since an extensive roof was chosen, the plants used will not directly absorb large portions of the water within twenty-four hours of the start of a storm. This means that the drainage system should be set up to hold the entire 5.05 inches of water. Assuming a minimum of two-inches of water to be held within the soil layer of the green roof, the inlet height of the drains are to be set at 3 inches above the top of the finished elevation of the green roof. Although this setup allows for the majority of the rainwater to be stored on the green roof, there are a number of factors that could lead to quantities of rainwater being wasted to the existing roof drains. First is the fact the green roof material between the saw-tooth structures will collect rainwater from the structures in addition to the area they cover. This means the between-structure green roof installation may see 10 inches of rainwater during the design storm. Also, if the plants do not absorb the rainwater as fast as designed, multiple storms within a few weeks could add additional rainwater loads to the roof. These design considerations lead to the necessity of keeping the existing roof drains in service.

Despite the fact the rainwater may be wasted under intense periods of storms, it is not recommended that the existing roof drain inlets be raised more than three inches above the installed green roof elevation. A three-inch water cover of the green roof plants shall be the maximum allowed amount to keep the plants from being damaged by flooding. The roofing material that has been developed for green roof is designed specifically for holding water. When choosing materials, the design engineer must be aware that this is important that the water-proofing layer can hold withhold the flooded area above from percolating through and into the building

When determining plant type, the cover of water during flood periods shall be kept in mind. Some plants will not be able to withstand rare three inches of water cover. The types of plants that were to be designed for, with the flooding in mind, are plants that absorb large portions of water without being damaged by flooding. The logistics of this design also demanded that these plants are hardy during dry periods, because the system was not designed with sprinkler systems to deliver water to the plants in times of drought.

Roof Type Determination

The type of roof that will be used for this project is the extensive green roof. It is important to use plants or a pre-grown modular roofing product that will require little or no supplemental irrigation, *i.e.* plants that can survive on just natural rainfall.

The roof should be as maintenance free as possible. This is also important because dried out; dying plants can be a fire hazard and may have insurance implications. The only type of roof that will support this type of vegetation is the extensive green roof. The vast majority of green roofs are extensive - shallower planting mediums. Intensive green roofs are the actual "roof gardens" that have planting medium as deep as two feet thick. These are almost never used on renovation projects because adding the structure to support that weight would be cost prohibitive.

Planting media for intensive green roofs are a foot deep at minimum, and have saturated weights ranging from 80 to 120 pounds per square foot. This type of roof is almost always used for new construction because it is difficult and expensive to reconstruct a roof to support this kind of load. Extensive green roofs, with a saturated weight of 12 to 50 pounds per square foot, are the most common. With planting media of 1 to 5 inches thick, most extensive green roofs are not designed for public access or to be walked on any more than a typical membrane roof would. Several modular extensive green roof products have emerged in the last few years that allow plants to be grown at the factory prior to actually being installed on a roof.

Another reason the extensive green roofs are the option to choose is because they are the least expensive option in the Life Cycle Analysis presented by Kosareo and Ries in their paper, Comparative Environmental Life Cycle Assessment of Green Roofs. The reason for this is because the intensive green roof must be designed for a larger load with shrubs, small trees, and other plantings. Also the intensive green roof requires much more soil and maintenance. The extensive roof requires less of a design load. The soil is much shallower, and there are not as many options for plantings. There should only be access to this type of roof for maintenance. The extensive roof tends to be less attractive visually than intensive green roofs, but less energy is needed for irrigation purposes. Finally, on many existing buildings, the extensive green roof will be the only option feasible due to Pre-existing design load constraints.

Plant Life for Extensive Green Roofs

Mosses are one of the most ideal types of plants which can be implemented on green roofs. They serve as the vegetative roof covers. Moss generally requires very less or no growing medium and virtually no maintenance. This reduces the additional weight on the roof caused by the growing medium. Moss can be grown on asphalt roofs with no growing medium and can easily survive no water conditions and recovers speedily from losing its green appearance without having great harm. Even a completely dried moss can regain its lush green colour in a single summer shower. It acts as the best substitute to conventional green roof plants as the former works well in shady conditions. Moss can absorb most of the rain water nearly 10 times its weight. Moss growth can reduce weed growth on roofs containing sub strata. Some of the different species of mosses used on green roofs are Frog(mood)Moss, Feather Moss, Fern Moss, Rock cap Moss, Cushion Moss, Hair cap Moss.

Growing grass on the roof is also one of the most ideal and easy procedures for implementing green roofs. The green roofs consisting of grass tops can effectively minimize the building temperature (indoor) and reduce the heat island effect caused due to hot sun and pollution from vehicles. The grass cover on green roof can be easily installed in the case of an extensive green roof. Grass does not require deep growing medium as its roots do not extend deep into the soil. The grass green roofs keep the substrate fixed and avoid it from slipping and lumping. It can retain rain water efficiently and can survive hot climatic conditions. It is not suitable for extreme temperature conditions as it develops dried or dead patches throughout. A grass covered green roof can provide habitat to insects. A newly installed grass green roof requires initial maintenance like watering and application of pesticides. In the case of extensive green roofs the installation can be done using readymade green roofs (turf mats) which consist of sheets of grown grass with a substrate, filter layer, drainage layer, and waterproofing layer. Grass roofs are not a good substitute for moss and succulents. The former requires regular maintenance when compared to the later two.

The growth of herbs on green roof helps to survive purposes like existence of green roofs and provide organic food and medicated plants. This can be a useful project for hotels. It can be one of the cheapest options apart from sedums. Some of the different types of herbs and plants that can be grown on green roofs are organic basil, parsley, sage, tarragon, chocolate mint, peppermint, spearmint, chives, and lemon balm, pear plants, cherry plants, apple plant, marjoram hot peppers and cayenne pepper plants. Tomatoes and edible flowers can be raised in wooden beds. Some of the widely grown herbs on green roofs are *Baptisia Tincitoria* (wild indigo), *Chrysopsis mariana* (shaggy golden aster), *Eupatorium hyssopifolium (*Hyssopleaf Thoroughwort*), Eurybia divaricata (*White Wood Aster*), Euthamia tenuifolium (*Button head Goldenrod*), and Monarda fistulosa (*Wild Bergamot*).*

Succulents include plant species like sedum and cactus. These are also called fat plants. Sedum and cactus are mostly suitable for arid climatic conditions and can with stand in dry soils or less moist soils. The succulents retain water in leaves, stems and roots which gives them a swollen appearance. The conditions like increased temperatures and less water availability can be easily sustained by these species.

The presence of hairy, waxy, spiny textures on the outer surface helps in reducing the air movement around the surface of the plant which results in less water loss. The species of plants belonging to the sedum type are a common choice for extensive green roof application as they able pollution filters. These roof plants can be grown in a layer of substrate of thickness 1.5 inches. The height of these species varies from 2 inches to 18 inches and spread through an area of 6 inches to 12 inches. Some of the preferred green roof plants are listed below:

Allium schoenoprasum, Delosperma nubigenum, Sedum album, Sedum album 'murale', *Sedum floriferum, Sedum kamtschaticum, Sedum reflexum, Sedum sexangulare, Sedum spurium, Talinum calycinum*, barrel cactus, beaver tail, fish hook, and claret cup cactus, all of the same family, work well on green roofs. They have properties that include drought resistance, attractive colored aesthetics when in bloom, resistance to high temperatures and sun light, great performance in shady conditions, wide spread, and strong winter survival.

LEED Certification Benefits

The green roof can have a very significant impact on energy use and environmental quality. Following the LEED standards, the benefits of green roof are easy to understand and categorise. Several LEED credits with a brief description of how they applied to the installation of a green roof on the North Engineering Building. Each credit is under a specific category type and numbered according to the credit's focus.

Table . LEED Accreditation Points to Accumulate

Sustainable Sites		
Credits	Description	Points
5.1 Reduced Site Disturbance	Local plants and vegetation used on the green roof for habitat creation and restoration.	1
5.2 Reduced Site Disturbance	The green roof is an open place for students and faculties.	1
6.1 Storm Water Management	Rainwater and runoff are collected on the roof.	1
7.1 Landscape & Exterior Designed to Reduce the Heat Islands	The green roof helped to reduce the heat island effect.	1
Water Efficiency		
Credits	Description	Points
1.1 Water Efficiency Landscaping	Water was not directed to storm sewers.	1
1.2 Water Efficiency Landscaping	No potable water was used for irrigation of plants on the green roof because native plants were chosen.	1
Energy and Atmosphere		
Credits	Description	Points
1 Optimize Energy Performance	Insulating properties of the green roof reduced energy demand.	

Some points could have been applied to the resource and material Section of LEED accreditation. However, recyclable and sustainable materials were not investigated for the green roof. Therefore, these could not be included in this portion of the study.

LIFE CYCLE ANALYSIS

Benefits of green roofs have been listed in general terms throughout this report. A life cycle analysis has the ability to quantify these specific benefits for installing a green roof on the North Engineering Building. In addition, life cycle analyses can help to clarify the effects of installing a green roof compared to the leaving the existing roof.

There are two major types of life cycle analyses (LCA) that have been defined and were available for use as part of this project. First, is the economic input–output (EIO) LCA. This analysis makes use of publicly available data to compute national economy wide assessments of the impact of the process studied using directly calculated or known associated costs. The second approach is a process-based LCA. This LCA allows for more specific results in a wider range of comparison categories. However, this type of analysis is more difficult to create. The analysis completed for this project used a combination of the two approaches. This strategy has become common within many of the more recent studies that developed life cycle analyses. The EIOLCA was used in this report when a direct number can be computed for costs. A process LCA was used when existing studies have previously determined methods for analysing specific qualities relating the life cycles of green roofs.

Since there is a very wide variety of effects that result from installing a green roof, and there is a wide range of benefits that can be realised from having a green roof installed, it was very important to define the boundaries. Boundaries define what is included in the LCA. The boundaries for the LCA done for this report included both installation costs of the green roof compared to the existing roof and long-term benefits of the green roof compared to the existing roof. The long-term benefits were evaluated using an estimate of a thirty-year life-cycle period. The installation processes included energy used and necessary plants and other materials for the manufacture and construction stages. The long-term benefits included energy savings due to additional insulation and benefits of air pollution removal. There are a number of long-term benefits that were not included due to a lack of numerical data. These benefits were not deemed to be less important and should also be considered when making a decision despite the lack of their inclusion in this LCA. These benefits ignored benefits included the urban heat effect, the increase of wildlife habitats, and the decrease of pollution from Run-off water. Other areas that were not within in the boundaries of the life cycle analysis included any operational or demolition cost or materials.

Another important consideration when performing the life cycle analysis was to determine functional units of comparison. The purpose of determining a functional unit is to allow direct comparison between different aspects of the LCA. In existing studies, square feet or square yards has commonly been used as a functional unit. This was difficult for this project, because the saw-toothed structures add a third dimension to the roof. Instead, the functional unit for this LCA was decided to be the entire North Engineering Building roof. All categories were normalised for installation of this green roof.

Before beginning the LCA, the comparison categories were determined to decide what aspects of this installation should be reviewed. This is especially difficult for all analyses involving green roofs because of the wide range of effects. However, specific categories used for comparison for this report were decided to be monetary costs, Global Warming Potential (GWP), Energy Savings, and Air Pollution Removal. The units of comparison for these categories were U.S. dollars, carbon dioxide equivalents (abbreviated GWP), terajoules (TJ), and kg of total pollutants, respectively. All of these were based on the functional unit of the entire roof.

The first step was determining the installation costs during manufacturing and construction phases. As listed in the report, $150 per square meter or $125 per square yard was used. The total area considered for green roof installation was 5150 square yards. However, only the half of the saw-toothed area (the roof portion between the saw-tooth structures) was planned to have green roof installed. This brought the actually area to be installed with green roof down to 3285 square yards. The installation of this green roof has negative benefits including the associated monetary cost, the use of electricity and the creation of GWP. At $125 per square yard, The monetary values associated with installing the green roof. Using an EIOLCA model, the associated numbers.

Table. Monetary values associated with the two options.

Cost of Installing a Green Roof	Cost of Leaving the Original Roof
$410,625	$0

Table. Energy used and Global Warming Potential Associated with two options

Installation Effects	Installation of a Green Roof	Leaving the Original Roof
GWP	176 GWP	0 GWP
Energy Used	2.09 TJ	0 TJ

The green roofs have been found to save on energy costs within a building due to their insulating potential. The studies have found conflicting numbers regarding the potential of a green roof's energy saving properties. These numbers vary because of differences in the type of green roof installed, the thickness of the layers of the green roof, the existing insulation installed on the building, the outdoor climate, the indoor air changes, and numerous other properties of the buildings.

For this life cycle analysis, it was established that the existing building has poor to moderate insulation, and the building energy savings for the type of green roof recommended was determined to be 37per cent. Since annual power bills cannot pinpoint the energy used solely to heat and cool the rooms that will have the green roof installed above them, an annual electricity use of 50 kWh per square yard per year for conventional roofs was assumed. Since 3285 square yards are to be covered, this yields an energy usage saving of 164,250 kWh per year.

Using a life period of thirty years at an average of $.11 per kWh (an estimated average price for electricity in Toledo over the next thirty years), a cost was developed and the EIOLCA model was used.

The final category investigated was air removal pollution. The numbers that were used for this category have been taken from a study done in Chicago. This means that the total kilograms of pollutants removed in Toledo may be slightly less due to an assumed lower level of air pollution in Toledo than

Chicago. However, this number was difficult to quantify and these assumed differences were be ignored. The number used for total air pollution removal was 84.6 kg per hectare per year or approximately 7 grams per square yard per year. The quantified number for air pollution removal for the life cycle period considered by installing 3285 square yards of green roof.

Table . Results of EIOLCA for energy use taking the additional insulation from the green roof into account

Energy Use Effects	Installation of Green Roof	Leaving the Original Roof
Cost	$341,476	$542,025
GWP	3590 GWP	5700 GWP
Energy Used	41.4 TJ	65.7 TJ

Table . Air pollutants removed by the two options

Energy Use Effects	Installation of Green Roof	Leaving the Original Roof
Air Pollution Removal	697 kg of total pollutants	0 kg of total pollutants

Finally, the overall results of the life cycle analysis have been summarised. Each of the comparison categories has been summarised for the entire system within the set boundaries.

This makes it possible to assemble a more informed decision based upon quantitative evidence for only the areas within the boundaries. It is necessary to realise that the positive and negative contributions from the green roof installation of multiple areas outside of the boundaries have not been included in this final comparison.

Table. Final LCA Comparison

Comparison Category	Installing a Green Roof	Allowing Existing Roof to Remain
Money	$752,101	$542,025
GWP	3766 GWP	5700 GWP
Energy Saved	43.5 TJ	65.7 TJ
Air Pollution Removal	697 kg of total pollutants	0 kg of total pollutants

Due to the fact that the functional unit used for this LCA was different than other LCAs completed for green roofs, comparisons to published data were not possible. In addition, no existing analysis was found that used the same comparison criteria.

However, the results of the life cycle analyses from the numerous studies referenced by this report have found comparable results. All LCAs found for green roofs installed on existing buildings are in agreement that the initial costs of installation are not easy to overcome through energy savings alone. However, the environmental benefits of green roofs found in these LCAs.

BENEFITS OF A WHOLE-BUILDING DESIGN APPROACH

Buildings are tremendously complex. Even the simplest facilities require a highly trained team of professionals for their design, construction, and, eventually, operation. Actually, it is amazing that many buildings are ever built, considering the necessary financing, land acquisition, code and jurisdictional requirements, extremely long time frames, budgets that seem like moving targets, design- and construction-team coordination, and numerous other roadblocks that occur along the way.

With all of this in mind, how do we create and operate buildings that reduce the impact on local, regional, and global ecosystems, as well as building occupants, while boosting the bottom line? This may seem like a daunting task, but the good news is a specific framework can help streamline the design and construction process.

DEFINING WHOLE-BUILDING DESIGN

Whole-building design refers to a design and construction technique that incorporates an "integrated design approach" and an "integrated team process."1 Basically, all of the elements of a building's design need to be considered, and all team members/design stakeholders need to be involved in new ways earlier in the process.

Collaboration among stakeholders is key. It is important not only to fulfill traditional design and construction roles, but include others not normally involved in elements of the design process, such as building operators and managers or perhaps even future general occupants. The objective is to foster communication among all of the parties that have technical or other operational input as early as possible to convey programming objectives and goals clearly while reducing or eliminating much of the last-minute coordination or budget pressures that occur later in more traditional building-delivery processes.

Getting stakeholders to think outside of their traditional roles and communicate design benefits and issues earlier may sound easy, but it can be challenging. Although both of these aspects of whole-building design are closely related, a larger challenge lies in integrating the design process itself. So, what does an integrated-design process look like? Because most industry professionals are more familiar with a traditional approach to design and construction, let's start with what integrated design does not cover.

Be Proactive, Not Reactive

Think back to the last project on which your firm worked. If you are an engineer, you probably were contacted by an owner, owner's representative, and/or architect and asked to provide a fee estimate and scope for a project that had been programmed and designed to a certain point. Certain design decisions already had been made, and several might have affected design elements you had planned on incorporating. Some of these decisions might have caused problems, and some of the tasks that needed to be accomplished as part of sound engineering practice no longer were possible. From the beginning, your design was in a reactive mode, and although you may be LEED or green-conscious, there probably was little opportunity to include those types of design elements.

What if your engineering firm was involved in the very first project meeting? You and all of your colleagues from the other necessary disciplines, as well as some new players representing the facility's future operators, could have been present. The meeting would have begun with some general information about the new project, followed by questions asked of everyone. Some of the things you heard may have given you ideas about how to approach design problems, and other participants may have come up with even more ideas based on your thoughts.

This is how a team approach fosters integrated design. Not only does this process tend to be more creative, this type of early collaboration allows most budget concerns to be avoided or incorporated into the project moving forward.

6

Bridges Engineering Structures and Construction Systems

BRIDGES

Bridge is a very important human invention. It is essential to our daily life and it facilitates a convenient transport. Mankind has learned how to cross rivers and valleys by bridges since the ancient time. Early bridges were made by trees, ropes or rocks, etc. The main function of a bridge is to improve our transportation to facilitate the travel. A bridge is like an aerial road that let our traffic be capable of striding across natural obstacles such as straits, rivers, valleys… In some big cities, bridges are used frequently as a crossover *to reduce traffic jam and raise transportation efficiency.*

The construction of a bridge focuses on bearing the weight (or gravity) of vehicles and humans on it. The load acts directly on the desk and then passes to the base. How can the load be passed to the ground efficiently? In the following, we are going to introduce different designs of bridges under different conditions.

CONNECTING THE DOTS IN BRIDGE CONSTRUCTION

The Mackinaw Bridge, the Millau Viaduct, the Beipanjiang River Railroad Bridge. Ever wonder how these bridges were assembled to cover such amazing and difficult spans? While there is no single construction method employed, there are several fundamental techniques typically used.

MAKING THE CONNECTION

Bridges of all types are vital links in infrastructure development. Beam, arch, suspension, truss, cable stayed, cantilever, truss-arch, and lattice truss, from pedestrian walkways to awe inspiring spans of spectacular gorges, each bridge has unique requirements and challenges during fabrication. The same bridge design may require completely different construction methods simply due to site restrictions and accessibility. One site may have unrestricted areas for stockpiling of materials, equipment staging, and traffic control detours while another site may have limited access and require unrestricted traffic flow during construction. The former may be able to use in place fabrication techniques while the latter would probably require prefabricated methods. So while there is no single construction method for bridge erection, there are several broad categories of fabrication techniques. These are:

- Falsework or staging: temporary framing and scaffolding to support structural elements during construction.
- Span-by-span: structural elements are fabricated in situ between supporting structures to create each span in a sequential process.
- Full span erection: fully prefabricated spans are constructed off site, transported to the bridge construction project, and installed whole between supporting structures.
- Balanced cantilever: segments are installed or fabricated in situ on opposing sides of a supporting pier until the span is complete.

Additionally, there may be more than one technique employed on a single project to achieve the most efficient construction process.

To facilitate these techniques at least one of two types of heavy equipment are usually present:

- Crane assist: structural elements are placed and temporarily held in position by one or more movable lifting cranes located on the ground, barges, or on the bridge itself as construction progresses.
- Gantry or launch girder: typically a horizontal steel framework supporting a track and carriage which hoists and transports structural elements along a bridge span, then travels horizontally to the next span.

STRESS CAN BE A GOOD THING

Another interesting implementation of modern bridge construction is accounting for pre and post completion loading. Since a bridge does not realize full structural strength until all the elements are connected, individual elements may be shifted out of position and/or pre-stressed prior to final connection. For example suspension and cable stayed bridges typically arc away from the true load line until the weight of the decking has been applied. For bridges using structural concrete a technique of post-tensioning is employed to ensure the structural element experiences primarily compressive loading. This is vital as concrete typically has poor tensile strength. Understanding these concepts is important not only during construction but also during bridge maintenance and inspection. A suspension bridge significantly deviating from its load line may indicate missing elements, or tensile microcracks in a concrete beam may indicate insufficient post-tensioning.

BRIDGE CONSTRUCTION SYSTEMS

From the bold impact of cable-stayed designs to the timeless grace of slender concrete arches, bridges are among the pinnacles of man's structural achievements.

Structural Systems has been designing, constructing, repairing and strengthening bridges since the 1960's, becoming an acknowledged leader in bridge works and bridge construction systems on an international level having completed projects throughout Australia, South East Asia, the Middle East, the United Kingdom and Africa. As part of the BBR Global Network of Experts, Structural Systems utilises the most advanced technology available.

The bridge construction systems service provided by SSL for bridge projects include:

- Engineering design
- Supply and installation of post-tensioning
- Supply and installation of cable stays
- Traveling formwork systems
- Structure packages
- Strengthening of existing bridges
- Bridge repairs and maintenance
- Monitoring and testing

BRIDGE TOWER CONSTRUCTION

The piles with diameter of 2,000 mm will be driven as deep as 77 m below ground, and on the island side the 120 auger piles will be piled under each of the two 320-m high bridge towers. The bridge towers will be concreted using custom self-climbing forms in pours of 4.5 m. A crane will be used on the first three pours, afterwards the formwork will start unaided moving through the hydraulic motion of modular elements. The pylons will be A-shaped, therefore, the use of standard forms will not be feasible. An individual set of forms has been arranged for each pylon.Transition between section types will be carried out at summer levels at the elevations of 66.26 m and 191.48 m. The use of self-climbing forms will make it possible to achieve better quality and decrease the time of construction of cast-in-situ reinforced concrete structures by half as much again.

Fig. Central Span Structure

The span structure has an aerodynamic cross section to resist squally wind loads. The deck cross section shape has been determined based on aerodynamic analysis and optimised following the results of experimental wind tunnel testing of the scaled model.

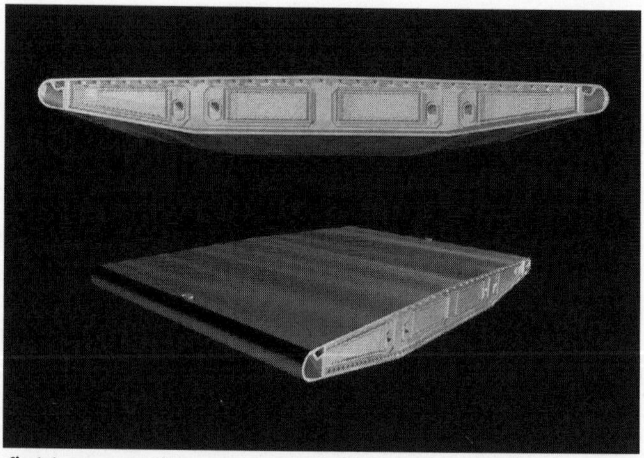

Welded field connections are used for longitudinal and transversal joints of the cap sheet of the orthotropic plate and lower ribbed plate. For joints of vertical walls of the blocks, longitudinal ribs, transversal beams and diaphragms, field connections are used provided by means of high-strength bolts.

Prefabricated sections for installation of the deck are supplied by barges to the erection site and hoisted by crane to 76 m elevation within dedicated intervals. Here, the sections will be linked and cable stays will be attached to them.

Fig. Cable-Stayed System

The cable stayed system assumes all static and dynamic loads on the bridge deck. Cable stays are provided with maximum possible protection not only against natural disasters, but also against other adverse effects.

The so-called "compact" PSS system has been implemented in the cable-stayed bridge deck; this advanced system differs by denser strands allocation in the sheath. Compact design of cable stays that employs sheaths of smaller diameter makes for wind load reduction by 25-30per cent. Moreover, the cost of materials for pylons, the stiffening girder and foundations decreases by 35-40per cent.

PSS cable stays consist of parallel strands of 15.7 mm diameter; every strand consists of 7 galvanised wires. Cable stays incorporate from 13 to 85 strands. The length of the shortest cable stay is 135.771 m, that of the longest is 579.83 m. The protective sheath of the cable stay is made of high-density polyethylene (HDPE) and has the following properties:

- UV resistance;
- resistance to local climate conditions of Vladivostok (temperature range from minus 40°C to plus 40°C) and environmental aggressiveness.

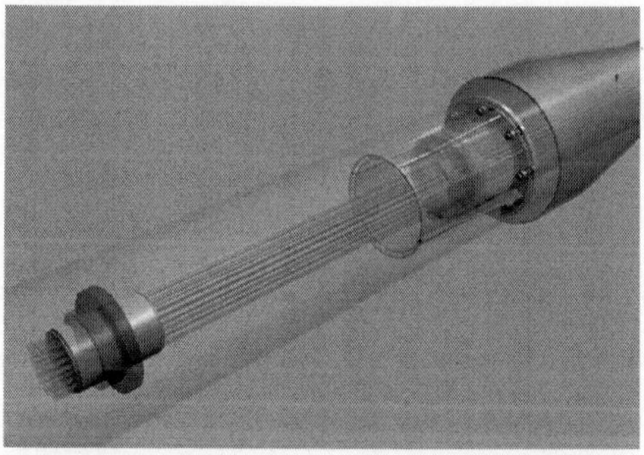

Fig. Construction of the bridge to Russky Island will be completed in Quarter II, 2012.

NEW CONCEPTS FOR ACCELERATED BRIDGE CONSTRUCTION

The evaluation process and the development steps for two new bridge concepts for accelerated steel bridge construction.

EVALUATION OF EXISTING SYSTEMS

All bridge systems are evaluated and assessed in light of the performance criteria. The performance criteria are grouped into four criteria. Each criterion consists of several categories with an assigned numerical rating. The total rating for each criterion is the summation of the numerical ratings for each category. The total score from all four criteria determines the ranking of the described systems. The four criteria with the associated sub-topics along with the numerical rating are presented below:

- Criteria I : Unit Configurations and Aesthetics (30per cent)
  - Aesthetics (15per cent)
  - Unit Configurations (15per cent)
  - Unit configuration is judged by the cost (5per cent) and the ease of fabrication (10per cent).
- Criteria II : Construction and Erection (25per cent)
  - Number of Joints (10per cent)
  - Transverse (5per cent)
  - Longitudinal (5per cent)
  - Joint Details (5per cent)
  - Ease of Erection (10per cent)
- Criteria III : Design Considerations (25per cent)
  - Fatigue Resistance (10per cent)
  - Joints Durability (5per cent)
  - Span Length (10per cent)
- Criteria IV : Future Maintenance (20 per cent)
  - Access and Inspection Efforts (10per cent)
  - Protective Coatings (10per cent)

The results of the evaluation for each stated criteria.

Table. Comparison of Unit Configurations and Aesthetics

Criteria I : Unit Configurations and Aesthetics				
Bridge Type	Aesthetics (15)	Unit Configurations		Total Score (30)
		Cost (5)	Ease of Fabrication (10)	
Temporary and permanent truss systems	7	4	10	21
Railroad Flatcar	10	4	8	24
Composite Space Truss	15	2	6	23
Steel Girders and Concrete Deck	12	5	9	26
Under-Slung Truss	7	3	7	17
Cold-Formed Steel Plate Box	13	3	7	23

Table. Comparison of Construction and Erection

Criteria II : Construction and Erection					
Bridge Type	Number of Joints		Joint Details (5)	Ease of Erection (10)	Total Score (25)
	Transverse (5)	Longitudinal (5)			
Temporary and permanent truss systems	3	3	3	9	18
Railroad Flatcar	5	5	5	9	24
Composite Space Truss	3	5	3	6	17
Steel Girders and Concrete Deck	5	5	4	9	23
Under-Slung Truss	5	5	4	7	21
Cold-Formed Steel Plate Box	5	5	4	8	22

Table. Comparison for Design Considerations

Criteria III : Design Flexibility and 75 Years Service Life				
Bridge Type	Fatigue Resistance (10)	Joints Durability (5)	Span Length (10)	Total Score (25)
Temporary and permanent truss systems	4	3	8	15
Railroad Flatcar	7	5	6	18
Composite Space Truss	7	5	9	21
Steel Girders and Concrete Deck	8	5	9	22
Under-Slung Truss	7	5	7	19
Cold-Formed Steel Plate Box	5	5	6	16

Table. Comparison of Future Maintenance

Criteria IV : Future Maintenance			
Bridge Type	Access and Inspection Efforts (10)	Protective Coatings (10)	Total Score (20)
Temporary and permanent truss systems	4	4	8
Railroad Flatcar	7	7	14
Composite Space Truss	7	9	16
Steel Girders and Concrete Deck	8	8	16
Under-Slung Truss	6	7	13
Cold-Formed Steel Plate Box	3	8	11

The overall numerical scoring of all bridge systems.

Table. Numerical Comparison of Existing Bridge Systems

Bridge Type	Unit Configurations and Aesthetics (30)	Design Flexibility and 75 Years Service Life (25)	Construction and Erection (25)	Future Maintenance (20)	Total Score (100)
Temporary and permanent truss systems	21	15	18	8	62
Railroad Flatcar	24	18	24	14	80
Composite Space Truss	23	21	17	16	77
Steel Girders and Concrete Deck	26	22	23	16	87
Under-Slung Truss	17	19	21	13	70
Cold-Formed Steel Plate Box	23	16	22	11	72

It can be seen from the results that the top three bridge systems are the Steel Girders and Concrete Deck, Railroad Flatcar and Composite Space Truss in the order of ranking. In order to closely examine each of these systems it is important to study the advantages and disadvantages of each system. The results from this comparison will help in formulating the new bridge concepts. Following is an examination of these three systems:

The rating values shown in the tables are subjective and are based on the experience of the research team with consultation with consulting engineers, bridge fabricators, and contractors. While rating of the established criteria might slightly vary, the conclusions will remain the same.

STEEL GIRDER/ PRECAST CONCRETE DECK SYSTEM

The steel girders with cast-in-place or precast decks are the most common elements in steel bridge construction.

Advantages:

- Improved efficiency with lighter steel beams (INVERSET brand system).
- Uses standard rolled shapes and welded plate girders.
- Economical/ average construction costs.
- Can be fabricated with exact camber and skew to meet existing site requirements.
- Top of deck can be textured for riding surface.
- Easy and rapid erection and construction.
- Cast-in-place concrete not required at joints.
- Suitable for use as continuous spans.
- Durable since cast in controlled conditions.

Disadvantages:

- In the case of the INVERSET type system, units are cast in "Upside-Down" Position and must be turned over in manufacturing plant after casting.
- Weight limit for transportation may limit the use in long spans.

Railroad Flatcar System

The railroad flatcar system has been used in the construction of bridges on low-volume roads.

The system relies on the availability of used and discarded railroad flatcars which limits the general use of this system. Therefore, this system is not promoted for widespread use; however, there are several aspects of the system, such as use of the flatcar as a stay-in-place form and details of the longitudinal joints, that can be implemented in a new general concept.

Advantages:

- Modular system.
- Easy and rapid erection and construction.
- Economical - about 2/3 cost of new bridge.
- Can be used on existing abutments and piers.
- Requires low maintenance if detailed properly.
- Effective, easily fabricated and constructed y longitudinal joints.

Disadvantages:

- Unknown fatigue resistance.
- Use only allowed on low-volume roads.

- Limited availability of usable flatcars.
- Supports must be at flatcar axle locations.
- Limited span lengths.
- Not suited for use as continuous spans.

Composite Space Truss System:

The composite space truss system scored the highest in aesthetics when compared to all other systems.

Advantages:

- High stiffness/ weight ratio.
- Alternate/ redundant load paths.
- Aesthetically pleasing appearance.
- Durable precast concrete deck slabs.
- Has great potential for modular design.
- Has potential for greater span lengths.
- Cast-in-place concrete not required at joints.
- Possible low maintenance if detailed properly.
- Suited for use as continuous spans.

Disadvantages:

- Current high cost due to complex fabrication.
- Complex welded "K and Y" joint fabrication.
- Large surface areas of steel tubing exposed.
- Requires lateral and vertical diaphragms at supports.
- Large number of transverse joints between panels.
- Possible need for riding surface preparation.
- Erection more difficult than other systems.
- Critical longitudinal connectivity of modules.

This system was closely examined by the research team due to its aesthetically pleasing appearance. The system could be modified to expand its potential.

A bridge concept consisting of several modular units tied together to form a bridge. Each modular unit consists of a fully prefabricated concrete deck supported by a space truss. The concrete deck can be prefabricated at the factory and transported to the bridge site or prefabricated close to the bridge and lifted into place.

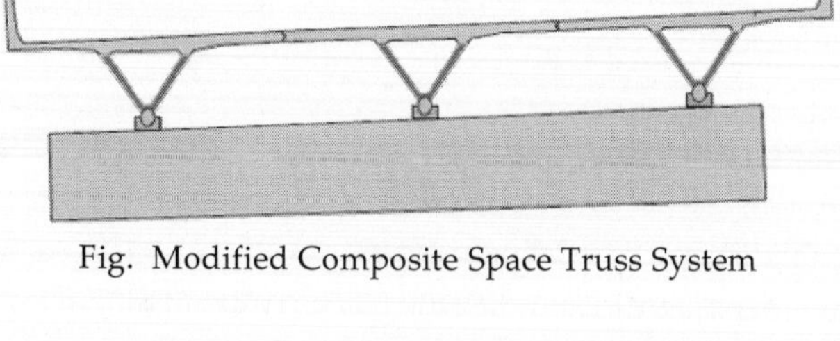

Fig. Modified Composite Space Truss System

The research team contacted several steel bridge manufacturers inquiring whether the tubular truss system can be easily manufactured with their existing equipments and fabrication techniques. All contacted fabricators voiced reservation on the practicality of this system and on their willingness to move into this type of fabrication. This fact alone will hinder the acceptance and widespread use of such system in the United States. Therefore, while at first glance this system offered the best option, the research team believes that it does not meet the objectives of the project.

NEW SYSTEMS DEVELOPMENT

The above systems evaluation indicates that a new system that combines the advantages of the steel girders with concrete decks and the railroad flatcar systems while minimizing or eliminating their disadvantages could offer the best options. The following criteria are used in establishing minimum performance limits for the new systems:

- Improved efficiency with lighter steel beams.
- Uses standard rolled shapes and welded plate girders.
- Economical/ average construction costs.
- Can be fabricated with exact camber and skew to meet existing site requirements.
- Easy and rapid erection and construction.
- Minimize number of joints.
- Suitable for use as continuous spans.
- Durability (Sufficient fatigue resistance to meet 75 years design life.)
- Extending maximum span lengths.
- Modular system.
- Can be used on new or existing abutments and piers.
- Requires low maintenance.
- High stiffness/ weight ratio.
- Alternate/ redundant load paths.
- Aesthetically pleasing appearance.
- Designed for HL-93 loading.

MODULAR CAST-IN-PLACE SYSTEM

Early in the development stage it was recognised that transportation limitation will play an important role in determining the modular unit configuration and the maximum span length. Also, the number and details of the transverse and longitudinal joints would have a marked effect on the durability of the new systems.

These limitations played an important part in the choice of the new concept. In addition to achieving the goal of accelerated construction, it is important to limit or eliminate any type of forming and shoring; in other words, all construction work needs to be performed "from the top".

The research team studied several concepts before arriving at the concept. The new concept combines the traditional steel girders and deck system with some features of the railroad flat car system. The basic components of the proposed system.

The new system relies on a wide interior unit with depth varying between 9 and 12 ft and a variable width exterior unit. These units are fabricated to create a self contained stay-in-place steel pan form.

A cold formed steel plate or corrugated metal form are welded to the top of the steel girders under factory controlled conditions followed by welding a welded wire mesh to the plate.

The steel pan depth is only 4 inches, the welded steel mesh is welded to the top of the steel pan. The depth of cast-in-place concrete deck will vary between 6 to 8 inches and will provide sufficient concrete cover to the steel mesh.

The welded wire mesh is designed to provide the necessary deck reinforcement. The stay-in-place steel pan form eliminates the need for forming or shoring and allows all construction work to be performed "from the top".

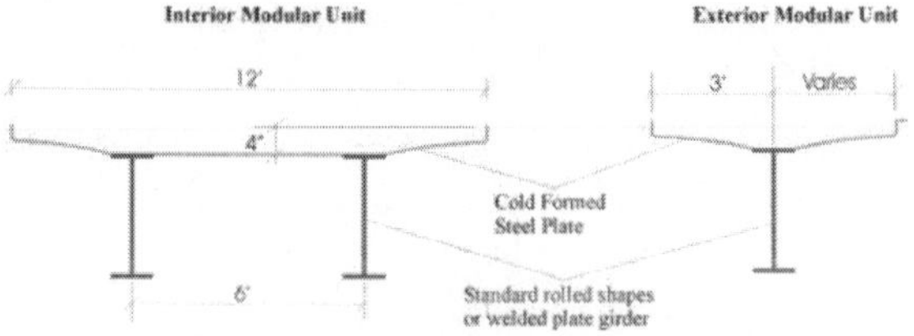

Fig. Exterior and Interior Units Details

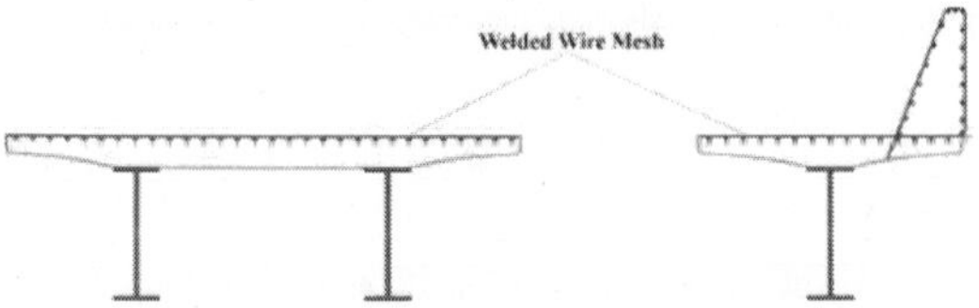

Fig. Adding Deck Reinforcement

The units are transported using traditional methods of transportation to the construction site and erected in place to form the bridge system. Each unit is tied to adjacent units using steel bolts spaced at predetermined distances. Once the units are assembled and tied together, sections of steel mesh for lap-splicing are placed on top of the longitudinal joints and tied to the welded steel mesh. The steel pan depth is only 4" and the wire mesh sections are welded at the manufacturer.

It should be pointed out, a unit width of 12 feet which is later used in the optimisation work, it is anticipated that the practical width limit will be between 9 to 10 feet to facilitate transportation of the units without requiring special permits.

The concrete is then placed in the stay-in-place steel pan forms to form the bridge deck. The use of fast setting concrete will allow traffic on the bridge within three days. The welded wire fabric placed on top of the joints are tied to the existing steel mesh using common wire ties. The cast-in-place concrete depth is 8" deep which places the welded wire fabric approximately at mid depth of the slab.

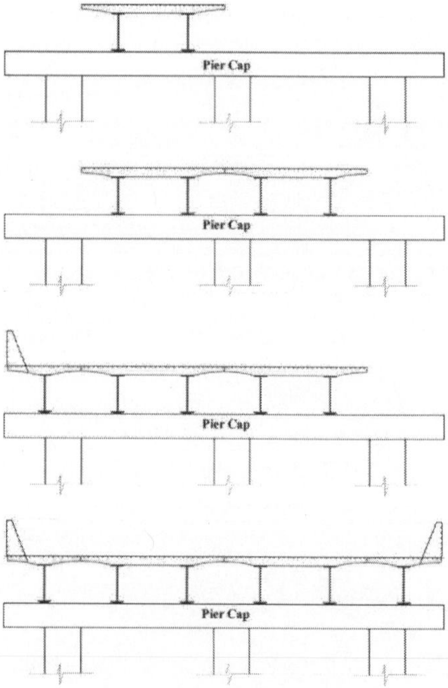

Fig. Erection and Assembly

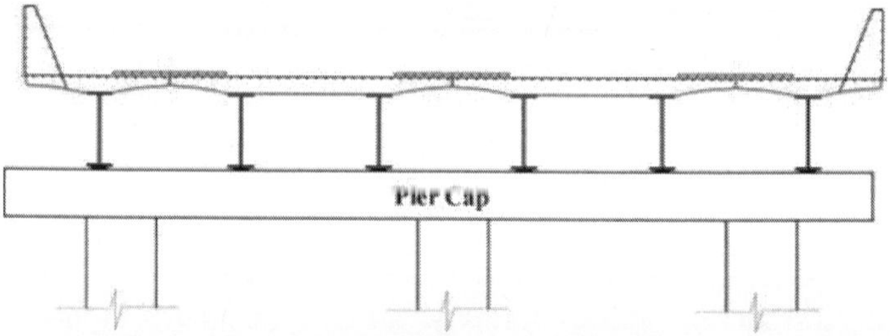

Fig. Installation of Steel Mesh Over Transverse Joints

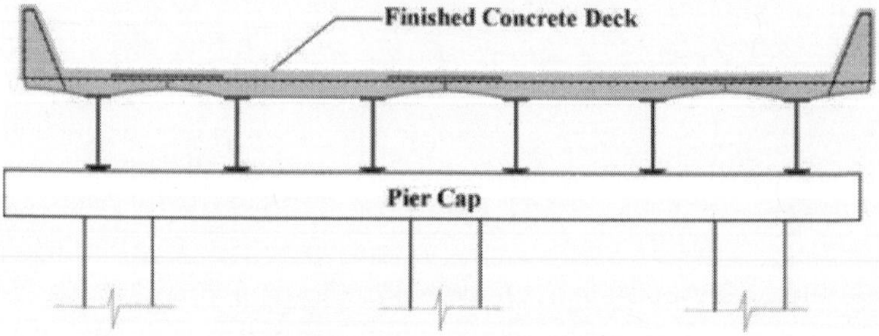

Fig. Placement of Concrete Deck

To facilitate construction and simplify the field work for forming and casting the exterior unit the exterior unit can be prefabricated and partially filled with concrete.

Block outs are fabricated during the construction of the unit to facilitate connecting and bolting the unit to adjacent interior units. The exterior unit is placed and tied to the interior unit. The welded wire fabric sections are then placed over the joints as earlier described. The cast in place concrete is then placed to form the deck.

If desired, continuity for live loads could be easily achieved by providing negative moment reinforcement over the support area prior to casting the deck concrete.

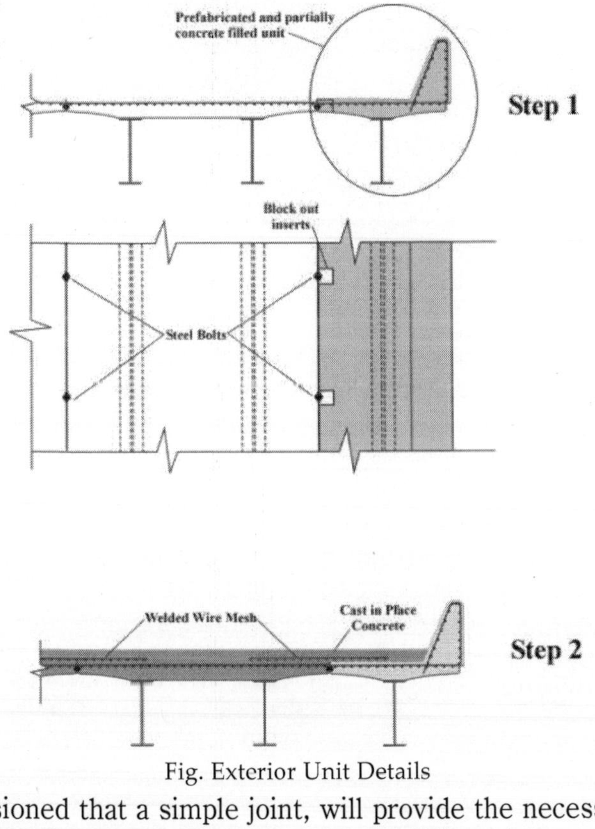

Fig. Exterior Unit Details

It is envisioned that a simple joint, will provide the necessary durability and performance of the proposed system. Joint details and the optimum number of bolts required to develop monolithic behaviour will be established through research and testing of the proposed modular bridge system.

The method described above will allow accelerated bridge construction, while eliminating all the construction joints, typical in modular bridge systems. Long term durability could be achieved through applying durable coatings to the interior and exterior of the modular units at the factory or through using high performance steel.

Fig. Transverse Joint Details

The proposed concept solves many of the problems associated with transverse and longitudinal joints. Though a very significant amount of concrete must still be poured to construct the bridge deck, the work can be performed without much intrusion to whatever is going on below the bridge. Once the units are up, the deck form plates provide a safe work platform while also adding some stability (note that these plate should probably be corrugated to ensure their own stability). The system achieves all minimum performance levels established during the conceptual development phase as will be seen from the optimisation process.

Modular Precast System

The system a modular precast system is proposed. The system consists of two steel girders spaced at 6 feet apart supporting a precast concrete deck. The total width of the modular precast unit is 12 feet. The width of the exterior unit is variable to allow simple customisation of the system. The concrete parapet is cast monolithically with the deck.

The modular units are transported to the job site using conventional transportation methods and erected in place to form the bridge structure. Once the units are assembled, locking bars are inserted into the longitudinal joints followed by casting the joint material. Durability of the joint could be achieved through proper detailing as well as the use of polymer concrete or epoxy as the joint material.

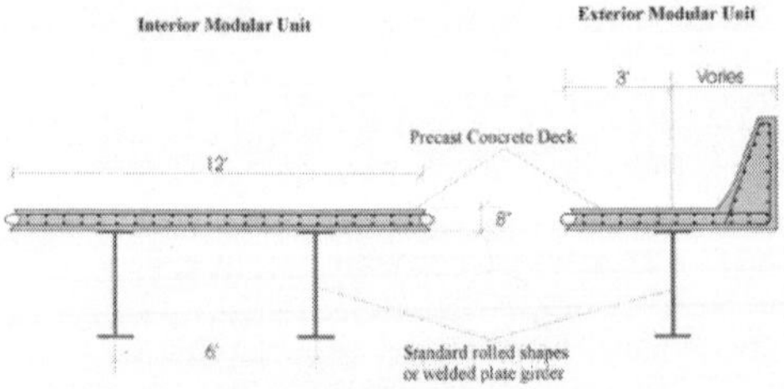

Fig. Precast Modular Bridge System

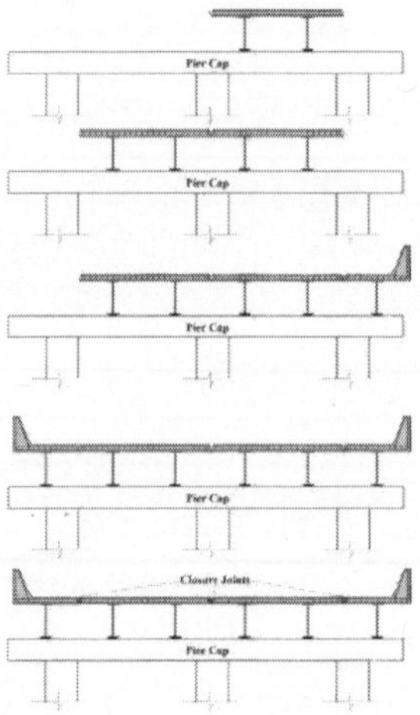

Fig. Assembly of Precast Modular Bridge System

Continuity for live loads could be achieved by extending the deck reinforcement at either side of the modular unit. Additional reinforcement is added and tied to the extend reinforcement followed by placement of cast-in-place concrete over the support.

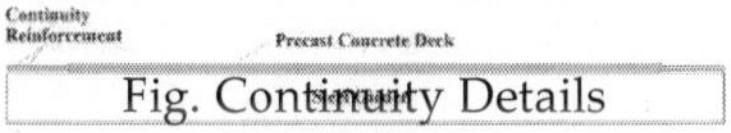

Fig. Continuity Details

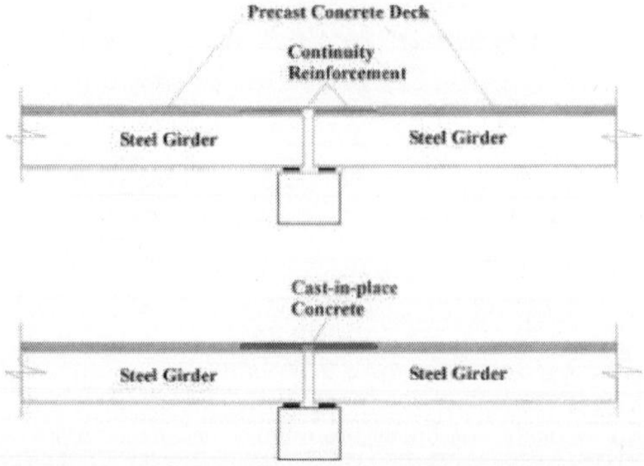

Figure : Precast Modular Bridge System Made Continuous

Transportation limitation will limit the maximum span lengths that could be achieved with this system. These limits are established in the optimisation analysis. It should be pointed out that up-side-down casting techniques or prestressing the steel are variations that could result in beneficial effects and should be considered in future research.

BRIDGE DESIGN AND CONSTRUCTION SYSTEM

Bridges are structures that connect two land masses separated by valleys or rivers, but they serve a greater purpose as they bridge the gap between different places, cultures, and societies. Discover how civil engineers construct bridges that win out over the deepest valleys and turbulent waters.

Millau Viaduct in France, a concrete structure supported by cables 270 meters high above the road level is a pinnacle of civil engineering achievement. It crosses over the valley of the River Tarn, and its overall length is 2460 meters. Across the globe, civil engineers have constructed bridges over such valleys and rivers, enabling the public to commute from one place to another efficiently. What we finally see is a structure carrying hundreds of vehicles from one point to another; however, the real story of a bridge is not that simple. Constructing a structure over rivers and valleys requires a lot of money, expertise, and patience. In this guide to bridge design and construction, we will get to know about the different aspects of bridge construction through many linked articles. Today, the study and advancement of bridge construction and design technology has diversified like never before, and we will discuss everything you should know about construction of bridges in this guide.

THE PLANNING AND DESIGN PHASE

First we plan and then we design and then we start with the construction of a bridge. However, the planning needs to be fail-safe because unless we know what to construct, how to construct, where to construct, we would not be able to erect a safe structure. Planning becomes extremely important when we have to deal with vehicles running at altitudes as high as 300 meters above ground level. Planning includes analysis of topography of the site, studying soil features, effect on wetlands, and surroundings.

DESIGN CONSIDERATIONS AND MATERIALS

Bridges have been constructed since ancient times. Simple bridges constructed from felled trees have been improved upon modern ones constructed by the modern light and strong materials. The latest computer technologies have significantly facilitated bridge design and bridge construction.

History Of Bridges

Bridge study has revealed that people have been carrying out bridge construction since humans first assembled into groups. The initial bridge design was basically felled trees that were utilized for moving over the ditches and rivers, and concrete bridges were rare. With the advance of civilization, techniques were discovered to use rocks, stones, mortar, and other materials for the creation of stronger and extended bridges. Subsequently, as the engineers and physicists advanced in the design, materials, and construction technology, modern materials like steel and aluminum were introduced for bridges.

Bridge Construction During 20th Century

The bridge construction skills progressed rapidly during the 20th century. At the end of the century, new techniques were developed that improved the design, strength, and durability of the bridges. Steel bridges were strongly riveted instead of the previous practice

of using bolts. Concrete bridges were being cast at the desired place, instead of being precast. Huge bridge elements made from bars and small sections were used, and not rolled as one part. Before the 1980s, the majority of bridge designs included expansion joints for decks, including expansion and fixed support bearings.

This technique was used to permit structural expansion and contraction. However, the expansion joints are likely to be filled with debris, and bearings often weaken over time. Thus, the structure is hardened, and maintenance requirements are increased. The bridge engineers explored methods to reduce this trouble, and finally the expansion joints and bearings were eliminated to develop a joint-less bridge. This type of bridge is constructed on a flexible foundation that may expand or contract with negligible trouble.

MODERN BRIDGE CONSTRUCTION TECHNIQUES

New technologies are expected to meet the challenging and varying requirements, and also offer options that will guide to innovative engineering and bridge construction standards. With the beginning of the new century, bridge construction is being revolutionized.Modern construction methods and the latest advanced materials are being evolved.

Construction technologies like post tensioning, reinforced ground walls, and soil freezing are being developed. Modern surveying techniques are being used that have facilitated the soil selection, and other design parameters, through the use of optical and infrared technology. Progress in the deck technology is creating lighter and stronger decks. Bearings, joints, and seismic elements have become more effective since advanced testing facilities have been introduced. Consistent, economical, fast, and programmed inspection systems will emerge.

NEW BRIDGE CONSTRUCTION MATERIALS

Materials with improved characteristics will be used that will make the bridge construction safe, durable, and reliable. Materials like high-performance concretes, polymer concretes, and plastics will be utilized. As the Fibre reinforced composites are becoming more tolerant towards temperature, they will be used extensively for bridge construction. Use of larger steel fibers will be used in the

tensioned members. The economics of future bridge construction will implement a simple design, with an increased interface between design, erection, and maintenance. Progressive study in modern superior materials and management techniques will facilitate the construction of durable structures that do not require extensive maintenance.

BRIDGE DESIGN CONSIDERATIONS IN AREAS OF SEISMIC ACTIVITY

Earthquake prone areas always exert great pressure on economy and social security. Safety is the most important aspect of bridge building for earthquake areas. Modern technology has now enabled engineers to build earthquake resistant bridges economically yet keeping the reliability factor intact.

Bridges do not just connect two far-off places; they serve a greater purpose. They minimize distance and help humans win over natural obstructions, rivers in particular. Building earthquake resistant *buildings* is important for human safety, but if a *bridge* is destroyed, connected places can be completely isolated from each other.

The 1989 San Francisco Bay Area earthquake suspended traffic for more than twenty hours because of the damage done to a bridge deck. In 2011, Japan suffered a major earthquake resulting in many bridge collapses, but the losses were minimized because the Japanese implemented earthquake resistant design practices. There have been many incidents in the US and other countries where post-earthquake studies have reported that a seismic design would have minimized the losses- or even absorbed the earthquake jolts.

The Design Considerations

While constructing bridges in earthquake prone areas, especially in the seismically active Zones III and IV of the Indian subcontinent, one thing we know in advance is that the chances of earthquakes occurring in such areas are very high. And that is the only thing we know because predicting earthquakes is not yet possible. While designing bridges in earthquake prone areas following things must be considered. Earthquake history of the region is a major factor that needs due attention from engineers. If a particular seismic zone experiences earthquakes of different magnitudes periodically, historical facts will help in predicting the most probable times of the year when an earthquake can happen. Obviously, you do not want to construct a bridge at the time when the chances of earthquake are high. Bridge building for earthquake areas needs extensive study of the earthquake history of the region.

Load Factors: Only dead and live loads are considered when designing bridges in non-earthquake zones. However, in quake prone areas, another force is of concern. This is called "seismic load," and it includes the forces on the bridge due to acceleration produced by earthquake jolts. How do we fit seismic load in our design? Historical facts help us to determine the seismic load factor in our designs. In case a region has no earthquake history, a minimum load is considered in design, which varies from zone to zone.

Seismic retrofitting is what modern technology has to offer. It not only makes old bridges resistant to earthquakes, but also helps in cost reduction. (Constructing a new bridge is always a time and resource consuming practice.) After a detailed seismic performance evaluation, retrofitting is carried out on old or weak bridges using methods such as friction damper systems, carbon Fibre plastic reinforcements, and external prestressing, as well as improving soil properties to keep a check on ground motion.

Seismic retrofitting is a relatively new technique but the Civil Engineering societies and associations across the world are promoting it because this is going to save money, time, and resources, without compromising the reliability and safety factor.

Minimizing Risk - The Bottomline

Japan, the pioneer of earthquake engineering, lost many lives and suffered huge property loss in 2011. Mastering technology is something that we humans can do, but the bottom line is that we cannot beat nature. This does not mean we should be pessimistic or passive; it means that we need to put our best foot forward and try to minimize the risk.

Earthquakes won't affect us as much if we are prepared, and the best preparation at this time is to implement earthquake resistant design practices and to make the best use of seismic retrofitting technology. An optimistic approach and staying updated with information related to seismic zones and seismic design considerations will definitely help to build better and more stable bridges.

BRIDGE CONSTRUCTION

Bridge construction has been improved tremendously with the advancement in science and technology.

Better and lighter materials are now available that can endure greater loads. The construction is now much faster due to the introduction of a variety of heavy construction equipment.

Bridge Construction Planning

Bridge construction tends to involve huge projects that encompass the utilization of skills related to several engineering disciplines including geology, civil, electrical, mechanical, and computer sciences. Therefore, integrating the efforts of all involved must be meticulous. The initial plans are prepared regarding the project, including the characteristics of the desired bridge, the site details, and the requirement of resources. The bridge design will be determined by the type of bridge being constructed.

The main types of the bridges are beam, arch, truss, cantilever, and suspension. The beam bridge is one of the popular types. Bridges can also be categorized by the planned use, like road and rail bridge, pedestrian pavement, material to be used like steel or concrete, and fixed or moveable. Moveable bridges are constructed when the ship height may be more than the bridge floor. In such situations, the road has the capability to be lifted or pivoted, to permit marine traffic movement under it.

Bridge Foundation

Construction of the foundations is the first step Towards building a bridge. This process involves detailed geotechnical investigations of the bridge site. The type of bridge foundation has to be selected, such as the well foundation, pile foundation, and the opened foundation. Each foundation is suitable for specific soil strata, and the desired bridge characteristics. The soil

characteristics will determine the load bearing capacity, and other important parameters. The superstructure is basically designed in accordance with the technical requirements, aesthetic reasons, and the construction methodology. Excavation required for the foundations may need to be executed to sizeable depths, involving hard ground, before the solid rocks are reached.

Engineering feats will be involved to avoid water, and prevent collapse of the diggings. Tunnels specifically may be subjected to sudden failures.

Bridge Construction Equipment

Heavy equipment will be used extensively during the bridge construction including bulldozers, excavators, asphalt mixers, formworks, and fabrication equipment. The construction and other equipment needs to be identified thoroughly, according to their capability and other desired functions. The foundation and the superstructure design will need to be considered. This expensive equipment shouldnot remain idle, and must be used cautiously to obtain optimum advantage.

Bridge Loads

Several loads act on a bridge, and the bridge is designed accordingly. Dynamic loads are particularly of prime significance. A bridge is designed to endure the normal vehicle loads, and other forces created due to winds and earthquakes. Several bridges have

collapsed due to high speed winds. Even if the wind speeds are reasonably low, the dynamic forces can become excessive for the bridge to resist. Initially, the bridge may vibrate violently, causing the bridge structure to fail at a few weak elements, or even damage the major components. Investigations conducted after bridge failures have revealed that the real forces on bridges that collapsed were significantly less compared to the loads for which the bridge was designed.

However, the oscillations created due to the winds were enough to cause the failure. Therefore, special reinforcement may be necessary for prevention against high speed winds and earthquakes. Thus, lighter materials are used that are arranged in suitable geometric structures, and it is ensured that the configuration is aerodynamically stable.

Testing of Bridges

Since bridge construction is an expensive project, it is essential that all necessary tests may be conducted prior to the actual construction. These tests and investigations can reveal the bridge Behaviour under different dynamic loads. Computer aided design

and testing are powerful tools that must be used to assist in the bridge design. Bridge design has benefited considerably due to the growth of computer Programmes. Such computer Programmes reveal immense information concerning the effect of different forces being applied on a bridge. Wind tunnels are being used extensively for the analysis of aircraft designs. Now these techniques are also being applied for bridge design examination. A wind tunnel is basically a space that is enclosed.

Air at a fast speed is moved through the bridge model. Likely design and structural defects can be discovered by photography and study of the air movement pattern over the model.

Once planning is compete, the focus shifts to which type of bridge to construct. Depending upon the outcome of planning studies, we arrive at a decision regarding the type of bridge to construct. A beam bridge is the simplest type of bridge whereas a truss bridge is reliable and economical at the same time. Different types of bridges are suitable for different types of climatic conditions. The following topics will guide you through much of the terminology and theory of bridge types and their construction methods.

TRUSS BRIDGE DESIGNS

Truss bridges are amongst a variety of bridge designs that are being used for road traffic. The basic shape of a truss bridge is like a right triangle, with the hypotenuse extending to the abutments. There are two similar trusses at each of the road sides.

Basics of Truss Bridges

Truss is a formation produced by triangular components, in accordance with the truss bridge drawings, and coupled at joints known as nodes. The triangular units forming the truss are slim and straight in form. The truss bridges consist of a grouping of triangles that are manufactured from straight and steel bars, according to the truss bridge designs. The solid arms of the triangle are extended from the pier sides. The diagonal steel tubes project from the bottom and top of each pier, and assist in holding the arms in the correct position. Trusses are organized as straight elements that are connected at the ends by hinges to develop a secure arrangement.

On application of loads on the truss joints, forces are communicated to the truss elements. The steel truss bridge members are in compression or tension. The trusses possess a high ratio of strength to weight, and therefore are useful for being employed in truss bridges. Trusses are also suitable for use in several other structures like roof supports and space stations. Amongst the modern bridges, truss bridges are considered to be included in the older kinds. The famous truss bridges are relatively inexpensive due to effective utilization of the bridge materials. The truss bridge designs are an important factor in architecture.

Truss Bridge Construction

Truss bridge construction is initiated with a detailed soil analysis to determine suitability for the bridge and traffic loads. The truss bridge is designed with either the road being at the truss bottom, or alternatively with the road being at the truss top. The former design has the bridge elements under compression, while the elements of the later design are mostly under tension. The bridge materials are selected and either welded or bolted, according to the requirement. The concrete is poured in the abutments, with the anchor bolts firmly inserted. Finally, the trusses are built, and the bridge is ready for use.

Truss Bridge Designs

A truss bridge is designed using Newton's laws of motion and incorporating pin joints. Pin joints are places at which straight elements of the truss formation meet. It is assumed that only the tensile and compressive forces operate on the truss elements. The truss bridge comprises vertical, horizontal, and the diagonal elements. The vertical members are under tension, while the horizontal elements are subjected to bending, shear, and tensile loads. The diagonal elements are under compressive forces.

Kinds of Truss Bridges

There are numerous forms of truss bridges that depend upon the topography and the purpose to be obtained by the bridge. The common types of truss bridges are as explained below:

Howe Truss

The Howe truss consists of diagonal and vertical elements. The diagonal elements slope towards the bridge center in an upward direction. The vertical elements are under tension. This is a rare type of truss bridge originally patented in 1840 by William Howe. Famous examples are the Jay Bridge in New York and the Sandy Creek Covered Bridge in Missouri

Lenticular Truss

This kind of bridge uses a lens-shape truss. An upper arch curves up and then down, and a lower one curves down and then up.

The two arches meet at the same end points. Examples of this bridge include the Royal Albert Bridge in UK and the Smithfield Street Bridge in Pennsylvania.

Bailey Bridge

This bridge is normally utilized to cover obstacles by the military tanks and other equipment. It can be constructed without the need of heavy tools or machinery.

K Truss

The bridge design is complex, and the construction requires extensive use of heavy equipment. The bridge elements under tension are reduced in number.

Bowstring Arch Truss

This bridge was patented in 1840 by S. Whipple. The main characteristic here is that the vertical loads on the thrust arches are transmitted along the arc path. At the end of the arch, the thrust is resolved into vertical and horizontal components.

Pegram Truss

In this bridge, the upper chords are all of equal length and the lower chords are longer than the corresponding upper chord. Because of this difference, each panel is not square.This truss design was patented by George H. Pegram in 1885. Only ten Pegram bridges remain in the United States, and seven of them are in Idaho.

BEAM BRIDGES

Beam bridges, also known as girder bridges, are one of the oldest and most common types of human construction. They are supported by land structures like river banks or by piers at both ends. Box girders or trusses may be used to strengthen the beam. Highway overpasses are normally beam bridges.

The History of Beam Bridges

Bridges have been constructed by human beings since the ancient times, with the initial design being extremely simple consisting of a tree placed across a stream or a river.

With the advancement of civilization, more useful methods were discovered in bridge building that were based on the utilization of rock, stone, mortar, and other construction materials to build bridges that were stronger and longer. Roman techniques for bridge building included the driving of wooden poles, at the intended location of the bridge columns, and then filling the column space with the construction materials.

The bridges built by the Romans were strong and of a uniform strength. After the industrial revolution, materials science developed rapidly with the introduction of materials with enhanced physical properties, and wrought iron was replaced with steel since it had a greater tensile strength. Presently, four main kinds of bridges are used, namely beam, arch, cantilever, and suspension.

Concrete Slab Beam Bridge

The Construction of Beam Bridges

The beam bridge, also known as a girder bridge, is a firm structure that is the simplest of all the bridge shapes. Both strong and economical, it is a solid structure comprised of a horizontal beam, being supported at each end by piers that endure the weight of the bridge and the vehicular traffic. Compressive and tensile forces act on a beam bridge, due to which a strong beam is essential to resist bending and twisting because of the heavy loads on the bridge. When traffic moves on a beam bridge, the load applied on the beam is transferred to the piers.

The top portion of the bridge, being under compression, is shortened, while the bottom portion, being under tension, is consequently stretched and lengthened. Trusses made of steel are used to support a beam, enabling dissipation of the compressive and tensile forces. In spite of the reinforcement by trusses, length is a limitation of a beam bridge due to the heavy bridge and truss weight. The span of a beam bridge is controlled by the beam size since the additional material used in tall beams can assist in the dissipation of tension and compression.

The Future of Beam Bridges

Extensive research is being conducted by several private enterprises and the state agencies to improve the construction techniques and materials used for beam bridges. The beam bridge design is oriented towards the achievement of light, strong, and long-lasting materials like reformulated concrete with high performance characteristics, Fibre reinforced composite materials, electro-chemical corrosion protection systems, and more precise study of materials. Modern beam bridges use pre-stressed concrete beams that combine the high tensile strength of steel and the superior compression properties of concrete, thus creating a strong and durable beam bridge.

Box girders are being used that are better designed to undertake twisting forces, and can make the spans longer, which is otherwise a limitation of beam bridges. The modern technique of the finite element analysis is used to obtain a better beam bridge design, with a meticulous analysis of the stress distribution, and the twisting and bending forces that may cause failure.

CHARACTERISTICS OF BEAM BRIDGES

A beam bridge is designed by the modern design techniques and the construction involves a thorough soil analysis, followed by the selection of bridge materials that can endure the bridge loads.

Beam Bridge Basics

Beam bridges are a common type of bridge among the truss bridges and the arch bridges. Wide flange rolled

form beam bridges are normally the most cost-effective type of beam construction that is used for beam bridges with short spans. The beams generally are located at fixed gaps by usingconstruction cranes, parallel to the traffic direction, between abutments or piers.

A concrete deck is placed on the main flange to provide lateral assistance against buckling.

Bridge deck forms are constructed according to the traffic load and other design characteristics. Diaphragms are fixed between the beams to provide additional reinforcement, and also facilitate lateral distribution of loads to the bridge beams.

Kinds Of Beam

Normally, the deck of a beam bridge is made of reinforced concrete or metal. Cross bracing is usually used between the bridge beams to increase the load bearing capacity of the beams.

Another technique of increasing the beam load capacity, with use of minimum web depth, is by the addition of haunches, located at the supported ends.

Generally, the middle section has is a customary form with a parallel flange. The angled or curved flanged ends are bolted, or riveted, by the use of joining plates. Since transportation of heavy and long beams to the bridge site is difficult, short convenient beam lengths are normally connected on site by using splice plates.

Advantages and Disadvantages of a Beam Bridge

Beam Bridge Advantages

- Beam bridges are helpful for short spans.
- Long distances are normally covered by placing the beams on piers.

Beam Bridge Disadvantages

- Beam bridges may be costly even for rather short spans, since expensive steel is required as a construction material. Concrete is also used as beam material, and is cheaper. However, concrete is comparatively not that strong to withstand the high tensile forces acting on the beams. Therefore, the concrete beams are normally reinforced by using steel mesh.
- When long spans are required to be covered, beam bridges are extremely expensive due to the piers required for holding the long beams. Building of the support piers may not always be possible due to the limitation of space.
- Bridge beams are likely to droop between the piers, due to the different bridge loads acting downwards. The forces acting upwards at the pier supports also influence the drooping effect. The sagging tendency is increased when the bridge span or load is increased.

Modern Beam Bridges

The sections of the current steel beam bridges are generally fabricated at appropriate locations, and then transported to the bridge construction site. Welding of the bridge elements is performed at the construction site. Subsequently, the sections are fixed with the supports. The shape of the boxes is designed to be trapezoid since this configuration facilitates passing of the wind, without damaging the bridge structure.

The maintenance of a steel beam bridge is extremely expensive. Concrete beam bridges being more economical, are therefore commonly used. The concrete beam is simple, and does not need a great deal of maintenance. Concrete beams include adequate steel for their reinforcement.

TYPES OF BRIDGES

Bridges have been used since ancient history though the technique of binding logs has been replaced by modern materials shaped into well-designed structures to resist applied loads. Here are the different types of bridges that exist in the world today

History Of Bridges

Different types of bridges have been constructed since ancient times. These bridges were si

of only trees tied together and used to cross rivers or channels. Gradually with the progression of civilization, other robust materials were used like stones and rocks to build longer bridges that used simple supports. Later the Romans built bridges that were of a uniform material strength by using cement consisting of lime, sand, rock, and water replacing natural stones that possessed varied strength. After the Roman era, bridges were built by using mortar and bricks. After the industrial revolution, wrought iron was initially used for the bridges, but since it did not possess adequate tensile strength to withstand large loads, it was later replaced by steel. In the present era of advanced technologies in all the disciplines of engineering, several computer Programmes have been developed that can design bridges with an accurate analysis of bridge performance under the different forces acting on it.

Types of Bridges

Bridges are designed in accordance with their planned use. For example the materials selected for trains, road traffic, or pedestrian paths are selected for their mechanical properties. The bridge design determines how tension, compression, shear, and torsion are distributed on the structure. The common bridges used are beam, cantilever, arch, suspension, truss, and floating bridges. They are different from each other in the materials used, construction techniques, shape, and span of the bridge. Beam bridges consist of reinforced steel girders that transfer the loads on the piers located at each end, while the cantilever bridge consist of two beams that are supported at one end only. Arch bridges are arch shaped with their weight being forced into the supports at each end and with construction materials that are light and posses high tensile strength. Similarly, other bridges are constructed according to their requirements.

Forces acting on Bridges

Tensile and compressive forces are present in all types of bridges. An efficient bridge design ensures that these forces are resisted without buckling, by either transfer of these forces to the areas of greater strength or dissipation of the forces over a larger area. The transfer of loads occurs in a suspension bridge, and dissipation of loads takes place in an arch bridge. The weight of the bridge is concentrated on the supports at the middle and the ends by the use of cables, girders, and arches. The supports rest on firm rocks, or reinforced boxes that are filled with concrete and sunk into the ground. The bridges are normally made from steel beams and steel since these materials possess the capability to resist the compressive and the tensile forces.

Future Bridges

Nations are spending huge amounts on bridge designs, construction, and maintenance because the reliability of highways and bridges has a significant effect on the economic growth of a country. The aim of future bridge designs is to construct bridges that are economical and long lasting by controlling corrosion, reducing structural maintenance, and having safe flexibility for modifications according to traffic demands and resistance to earthquakes, floods, overloads, and collisions. These goals can be achieved by using advanced materials, modern design and construction techniques, and superior inspection procedures.

The challenges can be achieved by a joint strategy of the highway department, research by universities, construction industry, heavy vehicle producers, and the users. Research activities need to be broadened and enhanced in the fields of steel bridge and fabrication technology, inspection procedures, and improved design by using modern computer Programmes. The engineers are exploring modern bridge design concepts, including new bridge shapes, use of superior materials, improved fatigue performance by ultrasonic impact treatment, concrete manufactured according to specific site conditions, and elimination of joints by using integral abutment construction, thus avoiding water leakage and corrosion.

WHAT ARE BEAM BRIDGES?

Simple beam bridges may use wood beams, but modern beam bridges use light and strong materials. Construction of a multi-beam bridge involves the use of a bridge crane, beam clamp for holding, and other tools. Bridge load rating is done to determine the loads that can be safely carried.

What is a Bridge?

A bridge is essentially a construction that is built to cover a road, valley, water body, or other natural obstacles to provide a route over the barrier.

Several bridge designs are used that depend upon their function and the soil conditions of the site for bridge construction. A bridge is described normally by its form of construction, like beam, truss, arch, etc. A bridge may also be characterized by the construction materials used, like concrete, stone, and metal. A bridge may have different types of spans that include simple, cantilever, continuous etc.

Beam bridges can be simple and made of wood beams. Heavy beams are carried by a bridge crane using a beam clamp to hold the beams. Bridge load rating is executed to establish the loads that can be safely carried by the beam bridges.

It is essential to calculate the bending moments in a beam to establish a safe design of the beam bridges.

Characteristics of Beam Bridges - Types of Beam Bridges

Beam bridges basically consist of beam that is laid across the piers or supports. The beam should possess the strength to bear the loads that are expected to be placed on it. These loads are borne by the bridge piers. The loads cause the beam top edge to be compressed, while the lower edge is being stretched and is under tension.

Existing beam bridges are formed by girders, normally box girders, trusses or I-beams, that are supported on strong piers.

- Box girders are stretched, box shaped elements that are more suitable to bear the twisting loads.
- Trusses consist of one or more triangular units connected at joints or nodes.
- I-beams are economical and simple to fabricate. They are simply beams with an I-shaped or H-shaped cross-section. The horizontal elements of the "I" design are flanges and the vertical is the web of the construction

Other beam bridges may be fabricated from concrete beams that are pre-stressed. These materials possess the steel characteristics to endure loads in tension, and concrete strength to bear the compressive loads.

The beam bridge's strength is largely influenced by the distance between the piers. Therefore, the beam bridges are normally not suitable for longer length, unless several such bridges are connected with each other.

The beam bridge's span is dependent upon the beam weight and the materials strength. As the bridge material thickens, its capacity to hold the loads increases. Therefore, the span could also be increased. However, a sturdy beam may become too heavy, and sag. The beam bridges can be supported by the utilization of trusses.

Beam Bridges Materials

With the advancement in technology, materials science has also advanced considerably. Beam bridges materials being used are strong, light, and durable. The advanced materials for bridge construction have good operational characteristics.

Such materials include reformulated concrete, composite materials that are reinforced with Fibre, steel, and pre-stressed materials. Pre-stressed concrete is well suited for beam bridge construction since it can endure excessive compressive stresses. Steel rods are fixed in the concrete that can bear the tensile loads. Furthermore, pre-stressed concrete is cheaper.

The current techniques include use of finite element analysis to improve the design of beam bridges. Distribution of stresses on different bridge elements is analyzed to ensure strong beam bridges that can endure the bridge loads. The beams should be held by piers at the ends to increase the bridge load bearing capacity. Concrete, steel, or stones are normally used for the construction of piers. Since stones and non-reinforced concrete are weak in tension, they are normally used for beam bridges that are designed for lighter loads.

SEISMIC RETROFIT, EARTHQUAKE RESISTANT TECHNIQUES

Bridge construction means taking care of various factors like soil liquefaction, earthquake sensitivity of the region, and working with different construction materials. Nowadays bridges are made earthquake resistant to minimize damage. Seismic retrofitting and base isolation techniques are modern age technologies that can effectively help in reducing the aftereffects of earthquakes.

MULTI-BEAM BRIDGE CONSTRUCTION

Multi-beam bridges are a very common form of construction for roads and other bridges that are required to span across distances less than one hundred meters. The beams are connected by an horizontal element that allows transport of men, materials, and vehicles.

Bridges Require Beams to Span the Gaps between Piers

Bridges are essentially meant to cross gaps caused by sudden changes in elevation that come in the way of a road, rail, or pipeline alignment because of rivers, streams, road and railway crossings, or valleys. The main structural components of a bridge are the piers or foundations and the structure that spans between the piers. This structure can be a supporting arrangement of beams or trusses and a deck which will allow the vehicles or materials to be transported over the gap being bridged.

Beam bridges are mainly used for smaller spans, and their spans will depend on the materials being used for the bridge construction. A number of beams may be used over adjacent piers and the top of the beams will be connected by the deck element. Such construction is referred to as multi-beam bridge construction. Materials used for the multi-beams may be steel, reinforced concrete, or pre-stressed concrete.

Quite often the depth of the beam becomes a limiting factor, especially in cases where headroom for areas below the bridge becomes a limiting factor. For larger spans the use of trusses or cables becomes more economical. Such truss or cable stayed bridges are also more aesthetically pleasing.

Requirements of Multi Beam Bridges

Multi-beam bridges have a number of beams covering the gaps between the piers of a bridge. The beams are required to be able to take the loads that are expected to be taken over the bridge. Such loads are dispersed over the various beams through the slab or deck element and can consist of the static loads of the vehicles and or materials, plus the dynamic loads caused by the moving loads. Beams of bridges also need to be designed for wind forces and vibrations that can be caused by the moving loads and wind. Vibrations are known to be very potent forces and have been known to have caused bridge collapses due to the harmonics set up by vibrations.

All the beams across a single span may be connected laterally with proper structural elements, and these can help to reduce vibration and lateral forces and add to the structural stability of the entire structure. Beams of bridges are made of I beams made of steel or concrete or even box girders made of the same materials. Box girders are said to be more resistant to twisting and therefore can resist vibrations and the effect of moving loads far more easily. When a load is transmitted from the deck to the beam, the top layers of the beam get compressed, but the lower edges of the beam are under tension. This tension is not suitable for concrete unless it has been properly reinforced. Pre-stressing is used for bridge beams and enables the top layer of the beam to be under tension and the bottom layer to be under compression. So when the loads are taken by the beams when placed in position, they find it easier to resist the forces imposed on them.

Construction of Multi Beam Bridges

Bridge building techniques have evolved vastly to allow the prefabrication of beams, whether of steel or concrete, in controlled conditions like a workshop or yard and then transporting the beams to the site by various methods. Beams can be launched from the piers or in case of river or sea crossings can be floated to site and raised onto the piers. Engineers have been quite often known to have used tidal forces to lift large spans on to piers without the use of cranes. It is also quite normal for very large spans being made up of short segments that are then joined together at site and techniques of pre stressing used to hold the segments together.

Stress and strain gauges are nowadays made part of beam and multi-beam bridge structures, and the information from these gauges is transmitted electronically to a monitoring site not necessarily at the bridge location. This allows engineers to monitor the health of bridges so that corrective action can be taken whenever needed.

STEEL TRUSSES

Overtaking the wood truss as a popular choice for structural design, the steel truss has become an important architectural and structural engineering element. Cold forming into curvilinear engineered steel trusses creates an astounding variety of architectural design.

History And Types Of Steel Trusses

As a relatively new building material, steel has become especially useful when incorporated into engineered steel trusses and steel plate connected timber trusses.

Attempts were made in the 17th and 18th centuries to use iron as reinforcing and structural members, especially in bridge design for railroads and other heavy load applications.

However, it became tragically apparent that structural iron quality control and design methods did not have the necessary reliability and cost factors required for widespread use.

With the development of the Bessemer process in 1858 the cost of manufacturing steel began to drop, and with the Linz-Donawitz process 100 years later steel began to seriously compete with wood as a structural framing element.

Electric arc furnaces enabled efficient steel recycling processes, further decreasing production costs. Finally, the cold rolled process for producing steel truss members now enables low cost production of steel floor trusses, steel roof trusses, entire steel truss buildings, and the elements used in the design of steel truss bridges.

All have now become cost effective, reliable, and safe alternatives to more traditional structural framing materials.

Design And Use Of Engineered Steel Trusses

The top chords typically bear loads directly, and the resulting tension and compression load distribution through the struts to the bottom chord accounts for the greater overall load bearing capabilitys of the wood truss.

A commonly observed use of the triangular steel truss system is for electric grid power line distribution towers, for example. In fact, any wood truss design can now be competitively manufactured using steel elements. The engineering benefits of doing so are lighter weight and better resistance to rot, insect damage, and splitting.

A significant architectural advantage of steel over wood is the ability to cold form arches and curves as chord members, further increasing the design versatility as well as engineering stiffness. By incorporating parallel chords much larger spans can be achieved for lightweight coverings and vertical constructions.

A parallel chord truss typically uses tubular steel elements and two or more parallel chords much longer that the webs, or struts:

Note that the parallel chords need not be straight but can be curvilinear, lending additional architectural variety to these structural elements. Knowing the expected live and dead loads, the bearing points, overhangs, and other structural factors, spans over 300' can be engineered with these steel trusses.

Arches are particularly useful with this application, creating large enclosed or semi-enclosed structures such as aggregate storage facilities, barns, manufacturing enclosures, workshops, etc.

Some Common Reasons for Concrete Cracks

- Uneven evaporation of moistures: The uneven evaporation of moisture from concrete causes various types of cracks. If the water evaporated from the surface of the freshly placed concrete faster than it is replaced by the bleed water, then plastic shrinkage cracks appear. The excess moisture left out after hydration is eventually evaporated and causes the concrete to shrink, but the shrinkage is resisted by subgrade, reinforcements, and the other parts of the concrete, causing drying cracks.

- Thermal Cracks: The inner temperature of the concrete increases due to the heat of hydration in the concrete. This temperature difference between the core and the surface causes thermal cracks.
- Corrosion: The volume of the rust is always greater than the volume of steel. So, if the reinforced steel becomes corroded, then it exerts force that causes concrete cracks.

Steps for Repairing Concrete Cracks

- Clean It: Remove all the loose dirt by using a chisel and light hammer or maybe by using only a chisel. Use a broom or small brush to clean the surfaces neatly. The idea is to make the crack surfaces hard and clean enough so that chemicals can be applied directly to the crack surfaces.
- Wash it: Clean the cracks with water and use washing powder. You can use a hose to spray the water. Remember that the cleaner the surface, the greater will be the strength of the repaired concrete.
- Dry it: Leave the clean crack surfaces for drying. You can use a cover during drying in case of a dirty environment.

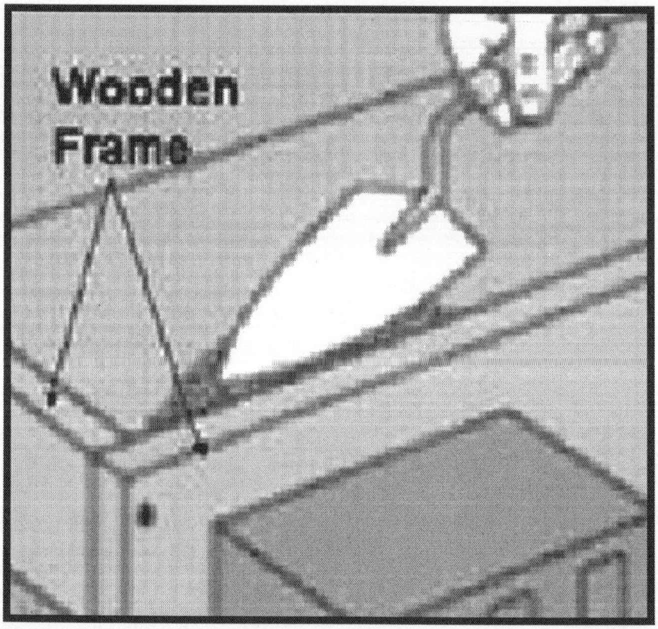

- For Small Cracks: Use the ready-made concrete repairing compounds. Prepare the compound as directed. Use a small chisel or a knife to prepare the crack. Pour the concrete repairing compound into the crack and use the chisel or knife to force the compound paste into the cracks. Once filled, smooth the surface and cover it up. Leave it for complete drying.

- For Large Cracks: Use readymade fortified concrete; there are various manufacturers available. The fortified concrete will be of powder form and you have to just apply water in it. Remember, once you add water the mixture should be consumed as quickly as possible, so make a small batch, sufficient enough for a single layer, at a time.

For the extra-large cracks or for the repair work at the sides of the concrete slab, you may have to use the wooden frames as shown below:

Remove the wooden frame as soon as the concrete become stable.

WOOD TRUSSES

For thousands years the use of wood trusses has been an important architectural and structural engineering practice. The concept has been considerably improved over the last 70 years to enable wood roof trusses and wood floor trusses to become common place construction components.

The Versatility of Wood Trusses

With hundreds of variations of a few standard designs, wood trusses have become some of the most versatile construction elements in use today. Wood floor trusses, wood roof trusses, open web wood trusses, wood bridge trusses, and a host of other applications are in use in many current construction applications.

While a solid wood beam is typically limited to spans of less than 25 feet and is usually not cost effective as an extended span structural element, they are still specified for certain high visibility architectural design applications. But by incorporating the chord and strut design, smaller dimension lumber can utilized which not only drives down costs but greatly increases the spanning and load bearing capabilities. Spans as long as 70 feet are possible with 2x6 lumber and loads exceeding 60 pounds per square feet can be achieved with properly engineered wood trusses.

Design of Wood Trusses

Whether prefabricated or built on site, the design of wood trusses used for structural purposes must performed properly. Engineered wood trusses are typically built according to custom specifications but "off the shelf" trusses are available for standard applications. The most common design is the triangular chord and strut configuration. This design evolved from Roman era engineering through the timber railroad bridges of the early 18th century. Following World War II it was realized that incorporating a flat metal connector plate along with stress testing and inspected lumber components yielded a much stronger, reliable, and cost effective truss.

The schematic of a typical triangular chord and strut wood roof truss design. The top chords typically bear loads directly, and the resulting tension and compression load distribution through the struts to the bottom chord accounts for the greater overall load bearing capabilities of the wood truss.

Each intersection will typically have a steel reinforcing plate with stamped barbs pressed into an overlapping position on both sides of the truss which serves to hold the members together. Many modifications of the same basic configuration are possible including, but not limited to, hip, scissors, vault, dual pitch, Polynesian, bow string, dual flat, cantilevered mansard, and almost any architectural design imaginable.

Wood floor trusses follow the same triangular load distributing design but have a long rectangular profile, where the top and bottom chords are parallel to each other. Struts can be spaced to accommodate utility chases, such as air conditioning and heating duct runs for example.

Knowing the expected live and dead loads, the bearing points, overhangs, and other structural factors will allow a manufacturer to calculate the required actual configuration. This process has evolved to the point of using automated manufacturing equipment to fabricate wood trusses to specifications, greatly decreasing Labour costs. And by specifying low deflection ratings, such as L/480 or smaller, structural vibrations can be reduced or eliminated as well yielding a very solid response to live load application.

ALL YOU NEED TO KNOW ABOUT CONCRETE EXPANSION JOINTS

All materials, including concrete expand or contract with the increase or decrease in temperatures. If suitable arrangements are not ensured for the expansion and contraction of concrete due to the temperature changes, cracks may occur.

- Concrete is not an elastic substance, and therefore it does not bend or stretch without failure. This concrete characteristic is useful and at times harmful on some occasions. Its high compressive strength and hardness make the concrete useful for applications in construction. However, concrete moves during expansion and shrinkage, due to which the structural elements shift slightly.

Why Expansion Joints in Concrete are Necessary

To prevent harmful effects due to concrete movement, several expansion joints are incorporated in concrete construction, including foundations, walls, roof expansion joints, and paving slabs. These joints need to be carefully designed, located, and installed.

The flexible expansion joints are basically designed to mitigate the flexural stresses. These stresses are produced due to the vertical movements of applications adjoining rigid foundation components, like columns or foundations. If a slab is positioned contiguously on surfaces exceeding one face, an expansion joint will be necessary to reduce stresses.

For example, if a slab is located between two structures, an expansion joint is essential adjoining the face of one building. Concrete sealer may be used for the filling of gaps produced by cracks.

Characteristics of Expansion Joints

Expansion joints permits thermal contraction and expansion without inducing stresses into the elements.

An expansion joint is designed to absorb safely the expansion and contraction of several construction materials, absorb vibrations, and permit soil movements due to earthquakes or ground settlement. The expansion joints are normally located between sections of bridges, paving slabs, railway tracks, and piping systems. Weather changes during the year produce temperature variations that cause expansion and contraction in concrete paving slabs, building faces, and pipelines.

The expansion joints are incorporated to endure the stresses. An expansion joint is simply a disconnection between segments of the same materials. In the concrete block construction, the expansion joints are expressed as control joints.

Road and Bridge Expansion Joints

Concrete expansion joints in roads and bridges are created in asphalt or concrete to allow material expansions and contractions due to the changes in temperatures, or due to road or bridge movement. The joints are cut at similar intervals in the structures to prevent development of cracking or splitting. Road expansion joints are normally sealed with hot tar, cold sealant, or compression sealants.

In addition, these joints are constructed by interlocking the metal edges. Reinforcing bars are also used to permit movement. In most of the cases, if movements are prevented, the roads or bridges may crack or buckle. The expansion joints allow controlled movement.

Installation of Expansion Joints

The site is prepared for the concrete pouring and the provisioning of the expansion joints in slabs. An individual expansion joint is created by the insertion of a flexible material that runs along the joint length. Suitable tools are used for making grooves in the poured concrete for placing of the joint materials. The depth of an expansion joint is usually one fourth of the slab thickness, or more if necessary. The expansion joint gap depends on the type of slab, like floating slab floor, vehicle pavement, sidewalk, or monolithic slab foundation. It is also influenced by the slab dimensions, type of concrete, and the reinforcing materials being used. Cracks in concrete may occur at the expansion joints due to improper concrete mix or curing. These conditions cause shrinkage between the expansion joints and cracks can be formed.

FUNDAMENTALS OF SEISMIC RETROFIT

Earthquake engineering is applied for the seismic retrofit of structures that involves alteration of existing structures to obtain resistance to seismic activity or soil failure due to earthquakes. Retrofit is now widely performed, especially in the active seismic zones.

What Is A Seismic Retrofit?

Seismic retrofit is a field of construction engineering that

focuses on the modification of existing structures to enhance their capability to resist earthquakes. Seismic retrofitting is achieved by the inclusion of structural improvements that may prevent the building, people, and the equipment from damage by seismic waves. In seismic zones, retrofitting may be essential for the bridges, overpasses, tunnels, and buildings, while the new construction would require compliance to seismic standards. Seismic retrofit may be executed on concrete masonry, unreinforced masonry, soft story, and concrete tilt-up construction. Soft story building is a multi-story building with abundant open space, and in concrete tilt-up construction, concrete is filled in the panels that form the walls of the structure. Since the concrete tilt-up walls are normally heavy, their seismic retrofit may be necessary. The motive for the concrete tilt-up retrofit is to prevent the separation of the roof from the building walls.

Categories Of Seismic Retrofit

There are several categories of seismic retrofit, and their nature depends upon the purpose for which retrofit is executed. The main form is the public safety retrofit that involves structural reinforcement to save human life (though some degree of injury is acceptable). Such retrofit is performed on structures that are not extremely expensive and for which a complete rebuild is not desired. A lower level retrofit may be selected to ensure earthquake endurance of the structure with some repairs needed after the earthquake. Extensive retrofit may be essential for the buildings that are important due to cultural, historical, or other reasons. Mostly, retrofit is performed on buildings that are extensively high, with an unsafe earthquake vibration frequency. Roads also have a priority in retrofit because their serviceability is essential during an earthquake to ensure the functioning of the emergency services.

Seismic Retrofit Measures

There are numerous techniques that are being used for the retrofit of structures,

and their selection is based on the objective of retrofit, type of structure, soil conditions, and the expenditures involved. Retrofitting of buildings may reduce the earthquake damage by suitable structural modifications, but no technique will completely eliminate the risk of seismic waves. Reinforcement of the buildings by the use of girders and trusses is normally implemented to make the buildings safe.

Base isolators are utilized to decouple the structure from the shaking ground, thus achieving seismic vibration control. Supplementary dampers minimize the structure resonant effects, increase the energy dissipation, and reduce the displacement of structures. Other retrofit techniques include the use of absorbers and baffles to make a building safe from an earthquake.

DESIGN PARAMETERS OF FOUNDATIONS IN SEISMIC ZONES

In seismic zones foundations are subjected to additional forces due to earthquakes. This will depend on the area and intensity of the earthquake. Design should take into consideration these forces.

Foundations in seismic zones are subjected to additional forces due to seismic waves. If the structure is not able to withstand these forces, it may collapse leading to serious catastrophes.

A foundation is the lowest part of a structure, which is normally below the ground level. Foundations are provided to transmit the load of the superstructure to the underlying soil.

It is necessary that the subsoil is able to withstand these loads. For this, the superimposed load should be lower than the safe bearing capacity of the soil. In case of foundations in seismic zones, additional loads are created due to the seismic vibrations. The design should take into account the additional forces.

General Considerations in the Seismic Design of Foundations

- Site investigations and determination of soil properties
- Details of geological and geotechnical environment
- Identification of loads - static and dynamic
- Type of foundation
- Safety verification as per building codes

Soil Investigations for Seismic Designs

For the design of foundations, the soil properties are to be ascertained. For this, both static and dynamic tests are to be conducted in the laboratory as well as in the field.

The important soil parameters to be analyzed are:

- Particle size distribution
- Relative density
- Shear modulus
- Damping factors

The important field tests to be conducted are:

- SPT Hammer Energy
- Pressure Meter Testing
- Shear wave velocity Measurement
- Cone Penetrometer Test
- Seismic Piezo cone Penetrometer

The purpose of these laboratory and field tests are to define soil deposit details, hydraulic conditions, soil index properties, static and dynamic stress-strain soil Behaviour.

Main Factors That Influence Site Effects

Seismological:
- Intensity and frequency characteristics of bed rocks in seismological environment
- Duration of bed rock motions

Geotechnical:
- Non-linear Behaviour of soils
- Elastic vibration characteristics of soils

Geometrical:
- Topography of underlying bedrock
- Non-horizontal soil deposit layering

Geological:
- Soil deposit thickness
- Type of under lying rock

Earthquake Characteristics

The main effect of an earthquake is the horizontal forces that are generated in the structure. The horizontal ground accelerations give a measure of this force. The granular soils get compacted due to the vibrations. This in turn causes the settlement of the ground surface. Another effect of the vibrations is liquefaction. The degree of liquefaction depends on the relative density of the soil, percentage of fines, depth of water table and the ground acceleration.

Effects of Earthquake
- The dynamic stress and induced pore water pressure may reduce the bearing capacity of the soil
- Loose granular soils are compacted by ground motion. This causes large subsidence of the ground surface.
- Compaction of loose granular soil may induce excess pore water pressure, which causes liquefaction of soil
- The vibration due to earthquake may cause structural damage

Measures to prevent Liquefaction

1. Compaction of loose soil
- with vibratory rollers
- compaction piles
- vibro floatation
- blasting

2. Grouting and chemical stabilization
3. Application of surcharge

Studying bridge failures is equally important because failure studies enable us to work on previous construction and execution mistakes. Bridges might fail because of human error or natural calamity, and analyzing failures helps a great deal to construct a better and safer structure in the future (because we can always learn from our mistakes). Here we have three very infamous engineering disasters that not only resulted in proprietarily damage, but also caused loss of human life.

Civil Engineering Wonders - The Best Bridges around Us

The Brooklyn Bridge was once considered the eighth wonder of the world; it is still called one of the greatest civil engineering feats of the 19th Century. America is proud of the Golden Gate Bridge in San Francisco, and it stands witness to the political and social changes the country has seen in the last century. Here we have articles mentioning the greatest civil engineering feats of the modern times, which include the Panama Canal bridges, the 126 meter high Bandra Worli Sea Link (BWSL) in India, and the tallest cable bridge in France. These are some of the great bridges around us that have benefited the masses and made the whole engineering fraternity proud as well. Are there any other topics concerning bridge design and construction that you would like to see covered at Bright Hub?

ARCH BRIDGE

Arch bridge is one of most ancient types of bridge design. Arcuate architectures were common in medieval time. By examining the arch bridge structure, we find that the vertical force acted on the desk would pass to the supporting points on both sides of the bridge through the designed arch-shaped curve. The following experiment simulates an arch bridge under pressure: Hold the ends of a long ruler with two hands and then put a book on the middle of the ruler. You will notice that the ruler deforms obviously. Then hold the ends of the long ruler again but haunch it up to simulate an arch bridge. Put a book on the middle of the ruler again. You will observe that the ruler has no obvious deformation.

As a result, the capacity on bearing the vertical force acting on a structure would increase if the structure is in an arch shape. The famous Sydney Harbour

Bridge in Australia is a typical example of arch bridge structure. Think carefully what happens when an arch bridge is under pressure! You may discover that the entire arch bridge structure is under pressure.

Therefore, materials that can bear high pressure, such as natural rocks, are the best materials for building arch bridges. Many old bridges are arch in shape for increasing the spans between the bridge piers. Anji Bridge, built at Zhaozhou in China in the 7th century, was once the bridge with the longest arch span in the world. This record had been kept for more than six hundred years until the 13th century.

Fig. Sydney Harbour Bridge

Fig. Anji Bridge

CONTINUOUS BRIDGE

Continuous bridge is another common type of bridges. Continuous bridge is usually built over shallow water areas or places where large ships cannot pass through. The *Island Eastern Corridor* in Hong Kong is a well-known example of continuous bridge design. If you observe its structure carefully, you may discover that the bridge desk was supported by many bridge piers.

Fig. The Island Eastern Corridor

CABLE-STAYED BRIDGE

Cable-stayed bridge also is one of the common types of bridges. Its special structural feature is that there are pylon/bridge towers built on the bridge pier with cables on both sides of tower.

To construct a cable-Stayed Bridge, we have to build a pylon/bridge tower on the bridge pier, and then suspend cables from both sides of the pylon/bridge tower to the bridge desk.

The load acting on the bridge is passed to the pylon/bridge tower through the cables and then to the ground through the pylon/bridge tower, so as to reduce the deformation extent of the bridge and the stress on the bridge. Steel cables with high tensile intensity are chosen as they have to sustain strong tension over a long period of time.

For examples, *Kap Shui Mun Bridge and Ting Kau Bridge in Hong Kong are designed in cable-stayed style using steel for making the bridge cables. The length of the cable-stayed bridge is usually between 160 and 900 meters.*

Fig. Kap Shui Mun Bridge

Fig. (Ting Kau Bridge)

SUSPENSION BRIDGE

The last kind of bridge we like to introduce is the suspension bridge. This type of bridge design is commonly used for striding across straits. The structural features of the suspension bridge are similar to those of the cable-stayed bridge: both designs are making use of the pylon/bridge tower. The bridge piers are connected to the bridge desk with a huge suspending cable and vertical steel cables. To construct a suspension bridge, we have to first build pylon/bridge towers on both sides of the bridge pier. Then the main suspending cables are hung on the towers.

Fig. (Golden Gate Bridge)

Then connections are made between the main cables and the bridge desk with vertical steel cables. The ends of the main cables are anchored to the ground by concrete to hold against the great tension acting on it. Golden Gate Bridge, located at San Francisco in California, is a famous suspension bridge. The recently constructed Tsing Ma Bridge, which connects between Tsing Yi and Lantau Island in Hong Kong, is also designed in suspension style with a surprisingly large span. In general, it is a very good bridge design in practice as the span of the suspension bridge can reach to more than 2000 metres.

We have summarised a lot of information on bridges. Is it helpful to you? Try to pay attention to the bridges around you in daily life and observe their structural features to see whether it matches what you have learnt in the above passages.

Fig. Tsing Ma Bridge

GROUND ANCHOR SYSTEMS

Ground Anchors is the generic term applied to systems which mechanically fix structures to the ground to resist load.

They are an essential part of modern foundation, abutment and excavation designs, as they can deliver hold-down forces of over 2000 tonnes each within a very compact package.

The design and installation of ground anchoring systems requires a high level of diligence if premature failure is to be avoided. Structural Systems is one of the world's leading ground anchor system specialists, having previously completed the largest permanent anchors ever used.

Equally vital is testing and monitoring of any anchor system both before and after it enters service.

Structural Systems has the resources to test installed ground anchors of any size for compliance with ISO standards, even in the most difficult locations!SSL provides design and installation services for ground anchors for all situations, including:

- Dam walls
- Retaining structures
- Cable-stay anchor-blocks
- Bridge abutments
- Slope stability
- Wind or seismic resistance
- Hydrostatic uplift

PRODUCTS, APPLICATIONS AND TECHNICAL INFORMATION.

Anchor Systems (Europe) Ltd supplies ground anchoring systems, for all forms of temporary and permanent works, and rapidly installed mini piles which have wide ranging applications. Operating throughout the UK and many parts of Europe, Anchor Systems has considerable expertise in geotechnical and structural stabilisation and offers a comprehensive technical support service including advice, design specification and site testing services as well as product installation, if required.

Duckbill mechanical ground anchors have extensive structural and groundwork applications and have proved themselves to be reliable in practically all displaceable ground conditions. With rapid installation, Duckbills provide an efficient and cost-effective means of stabilising stone and masonry structures as well as slopes and embankments, comprising virtually any substrate, and have been successfully employed on a diverse array of projects around the world.

Helical anchors are screwed into the ground to provide a quick, simple and reliable fixing point to which chains can be attached to secure items such as plant, equipment, motorcycles and bicycles. They can also be used for securing guyed structures, trees, vines and fencing and can be unscrewed for re use in other locations.

Excalibur anchor bolts, with their unique twin helical thread, are an innovative and versatile means of rapidly securing into bricks, blocks, wood and all grades of structural concrete. With a multitude of uses, such as securing machinery, benches, balustrades, fire escapes and crash barriers, they are rapidly installed without any need for additional fixings or special tools - to save both time and money.Soil Nails Anchor Systems supplies a comprehensive range of high performance soil nailing systems for the reliable and cost-effective stabilisation of embankments, cuttings, vertical walls, coastal defences and temporary support for basement works.Anchor Posts Anchor Systems (Europe) Ltd specialises in geotechnical and structural stabilisation. It offers a comprehensive advice, design, specification and site testing service for it's range of products., which include Duckbill mechanical ground anchors, soil nails, helical anchors, anchor bolts and the innovative new anchor post system.

Backstay Anchors Formed in 1995, Anchor Systems (Europe) Ltd supplies the well proven and versatile range of Duckbill mechanical ground anchor systems throughout the UK and Europe.

Duckbill Tree Kits are world renowned and have been long established as the leading anchoring system for semi mature tree plantings as well as those that require securing during or after storms by way of guying. The traditional range of anchor kits has been the rootball, deadman and overhead guys however, many other systems are available, custom made to specific requirements; with anchors capable of achieving 300kN, impossible is nothing. With a history of quality and service behind us, we can and will supply next day the systems you require as well as a range of installation equipment if required.

Sock Anchors are a mechanical and chemical anchor system for stabilising structures such as bridge abutments, retaining walls and embankments. The system comprises a high strength steel Rib Bar surrounded by a woven elastic polyester Grout Sock, sealed at both ends, which is inserted into a pre-drilled hole in the structure to be stabilised and pressure grouted to form a strong mechanical and chemical reinforcement.

CLASSIFICATION OF GROUND ANCHORS

Ground anchors are classified according to their service life, purpose, installation procedures and method of load transfer from the anchor to the ground. An anchor with a service life greater than 24 months is generally considered permanent. Permanent anchors shall always have some type of corrosion protection system based on the service life of the structure, the known aggressivity of the environment and corrosive properties of the soil and consequences of tendon failure.

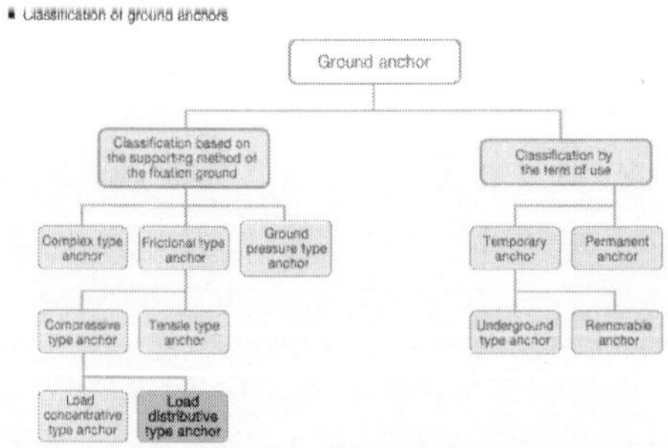

Fig. Comparison by the Characteristics of Each Anchor Type

Also, anchors can be classified into frictional type anchors that are supported by the friction of the grout and the ground, ground pressure type anchors that acquire anchoring force with the passive resistance of the ground using ground pressure boards or piles, and complex type anchors that are a combination of the above two types, based on the supporting method of the fixation ground. Frictional type anchors can also be classified into tensile type anchors and compressive type anchors based on the load application method to the grout. Lastly, compressive type anchors can be classified into load concentrative type anchors and load distributive type anchors depending on the distribution of the load.

Load Concentrative Tension Type Anchor

When stress is applied to tension type anchor, load transfer occurs to bond length through adhesion of steel strand and grout. Due to load concentration, the parts of tension type anchor attached with steel strand and grout become unzipped and this leads to crack and load reduction. In addition, tension type anchor has the weakness of progressive debonding and time-dependent load reduction (creep) occurrence when friction of load concentration zone exceeds the extreme skin friction of the target ground. The tension at the earlier phase displays the state as of 1.

Then, as the parts attached with steel strand and grout become unzipped, it changes into the state as of 2. The relatively concentrated skin friction of anchor becomes higher than the allowed value between ground and grout body to progress into the state as of 3.

Accordingly, load reduction takes place. The drawbacks of tensile type anchors are that a progressive destruction occurs due to the jacking crack in the grout and creeps due to load concentration, greatly reducing the load. Therefore, as in the vicinity friction distribution graph, the load transference distribution is as shown in curve 1) at the initial point when the load is applied, which changes into curve 3) due to the above mentioned reasons with the progress of time, reducing the load.

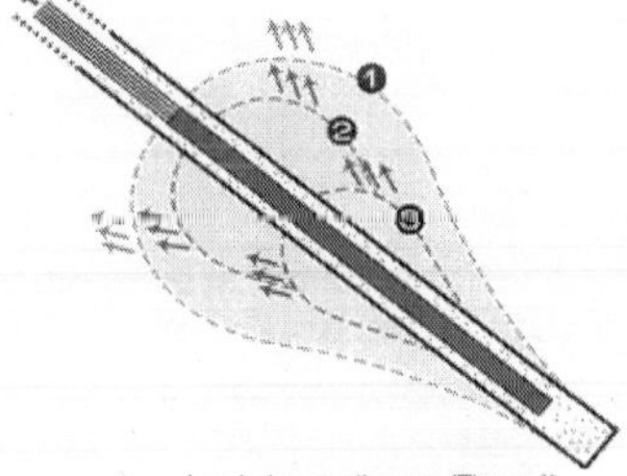

Load change diagram (Figure 1)

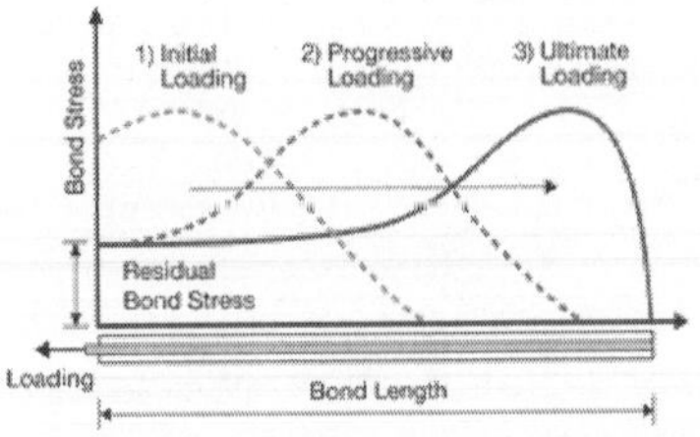

Vicinity friction distribution graph (Figure A)

Load Concentrative Compression Type Anchor

Compression type anchors consist of an unbonded polyethylene (PE)-coated steel strand which transfers the jacking force/ load directly to a structural element located at the distal end of the anchor. Unlike the tension type anchors, the grout body for compression type anchors is loaded in compression which is capable of securing much higher loads.

However, due to the concentrative design of these anchors, the use of high-strength grout is frequently required to secure the jacking forces at the distal end. Also, it is often difficult to secure concentrative anchorage force in weak soils. Similar to the tension type anchors, compression type anchors are subject to the occurrence of progressive debonding and time-dependent load reduction (creep) as displayed in state '$ as shown in. In this case the friction required to secure the concentrated load exceeds that of the skin friction for that zone. This effect causes grout debonding and loss of soil confinement pressure resulting in load reduction as displayed in states a$andb$.

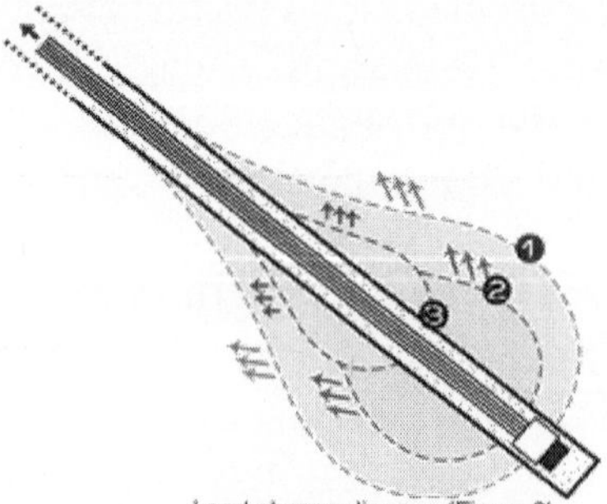

Load change diagram (Figure 2)

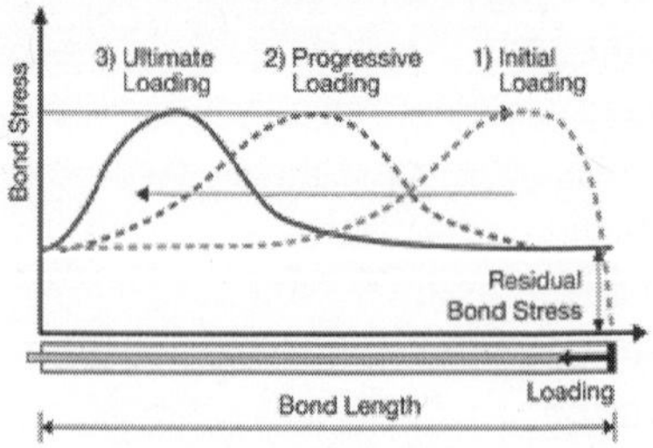

Vicinity friction distribution graph (Figure B)

Load Distributive Tension/ Compression Type Anchor

The high stresses from tension and conventional compression type anchors transfer concentrated loads to the soil and grout body which can become overstressed resulting in failure. Therefore, load distributive compression type anchors have been developed and are being used, which uniformly distribute the anchor load to the grout body and soil along the theoretical length of the bond zone. In addition the grout strength requirements are reduced as well as applied eccentricity.

As a result high loads can be achieved even in normal soil condition. Recently, load distributive tension type anchors have been developed which are capable of securing stable loads in even relatively weak soils such as clay and silts. These anchors do not require high strength grout and have low eccentricity as well. The use of load distributive anchors results in a more uniform distribution of the anchor force to the soil. Therefore, load reduction and creep are minimized, enabling the anchor to maintain initial design load.

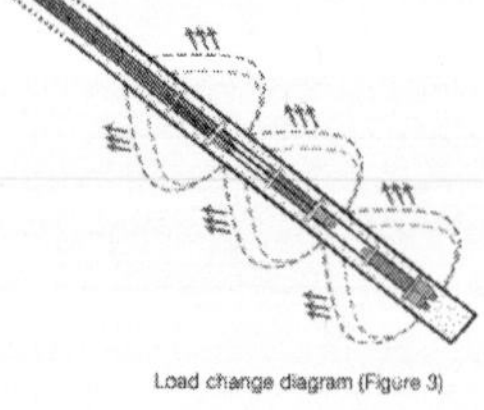

Load change diagram (Figure 3)

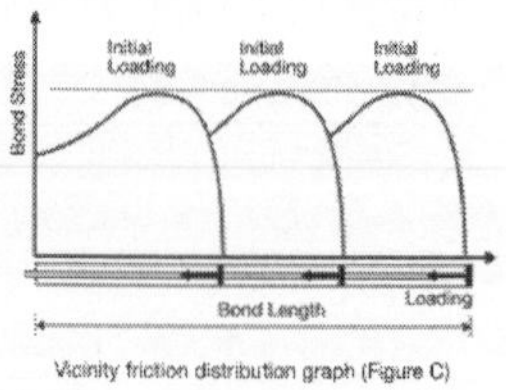

Vicinity friction distribution graph (Figure C)

We design and construct both permanent and temporary anchors using either multi-strand or stress bar in accordance with internationally recognised codes that have been adopted in New Zealand which include; British Standard Code of practice for Ground Anchorages BS8081:1989, Execution of special geotechnical work – Ground Anchors BS EN 1537:2000, FIP Design and construction of prestressed ground anchorages April 1996 and US Federal Highway Administration Geotechnical Engineering Circular No. 4 Ground Anchors and Anchored Systems June 1999.

We have an extensive database of ground anchor testing and geotechnical ultimate bond ruptures in a wide range of materials nationwide. However, in any ground anchoring project, site specific geotechnical investigations for the construction of the ground anchors are critical to minimise the risk profile for all parties. Access to this information means we are better able to qualify potential drilling difficulties and any potential for loss of grout from the drill hole during the anchor installation.

For multi-strand anchors, the strands are run through our specialist greasing and sheathing machine. This machine parts the individual wires of the strands followed by immersion into a grease bath before completely encapsulating the strand in the outer sheathing to ensure no voids are present.

We also provide mini-piles (drilled and grouted micro-piles up to 300mm diameter) which are structural supporting members that improve the stability and load bearing capacity of structures. These are constructed by drilling small diameter holes and constructing piles with high tensile steel and high compressive strength grouts to achieve tremendous load bearing capacities.

We utilise the technique of high-pressure (1000psi) post-grouting in weak compressible soils to significantly improve the bond capacity, and where required, fabric socks are used to ensure containment of grout within the drill hole thus ensuring full bond potential is realised.

Grouting Services, through its partnership with SAMWOO of Korea, offers world-leading ground anchor technology that includes:

- Removable, compressive, distributive anchors (SW-RCD)
- Permanent, tensile, distributive anchors (SW-SMART)
- Permanent, compressive, distributive anchors (SW-PCD)
- Permanent, tensile, frictional anchors (SW-PTF)

The load distributive compression (and tension) type removable anchors provide significant advantages to building owners as once the anchors are removed at the end of the project construction, there are no obstructions left in the ground that will conflict with any future developments.

BRIDGE ANALYSIS MODELS

Bridge Analysis Models (BAM's) are idealisations of a bridge used to compute the structural responses. The BAM in QconBridge II is used to analyse a bridge structure using the simplified analysis method as described in the LRFD Specifications.

This method analyses the bridge as a collection of separate, but related, *Longitudinal Bridge Analysis Models (LBAM's)* for analysing the longitudinal behaviour of each girderline, and *Transverse Bridge Analysis Models (TBAM's)* for modeling the transverse behaviour of each pier in the bridge. Once created, each longitudinal model is treated independently to model a particular girderline.

Then, the results from these longitudinal analyses may be used to apply superstructure loads to the transverse models. Hence, the BAM is not a model it its own right: It represents a collection that manages the interdependencies between the LBAM's and TBAM's.

Longitudinal Bridge Analysis Models are used to perform a single girderline analysis of the bridge superstructure. LBAM's are an idealized representation of a bridge using beam members, distribution factors, and idealized loads. LBAM's consist only of the information required to analyse a bridge. They do not contain product-level information about a real bridge (*e.g.*, you can't tell what type of girders they are made from, or what materials were used). LBAM's in QConBridge II can be used to idealize and analyse any type of bridge that can be modeled using the LRFD simplified (distribution factor-based) approach.

Transverse Bridge Analysis Models are used to analyse piers for live and dead load responses. TBAM's are an idealized representation of a pier using beam members and idealized loads. Loading input for a TBAM is typically synthesised from the results from one or more longitudinal analyses.

MODEL GENERATION

In order to perform a structural analysis on a Product Model-based bridge, there must a successful mapping from the product world to the analytical world. This is called Model Generation. QConBridge II uses a fixed set of assumptions for automatically generating BAM's from Product Models.

PRODUCT MODELING OF GENERAL BRIDGES

General Layout of All Product Model Bridges

A considerable amount of detail can be required to describe a Product Model of a specific type of bridge. However, in a computer programme, it is convenient to describe all types of bridges in a generalised and consistent manner. In QConBridge II, we take the following approach to describe all types of bridges in terms of generalised components:

1. Describe the roadway geometry. This defines the general orientation of the superstructure and provides a reference geometry for all bridge elements.

2. Describe the locations of supports (abutments and piers) with respect to the roadway geometry.
3. Describe abutment and pier details including the layout of columns and descriptions of the crossbeam.
4. Describe the width of the roadway and the geometry of sidewalks and barriers.
5. Layout girder lines and select girder types
6. Describe the connections between the girders (superstructure) and the substructure.
7. Describe other details such as overlays and reinforcement.

Although some items depend on others (*e.g.*, a roadway must exist in order to place piers on it), the sequence above is not fixed, and can be changed to suit your needs. Based on the above description, we can sketch an idealised Product Model bridge.

Also note that although the resembles a certain type of bridge, it is meant to be a general representation: all Product Model bridge types in QConBridge II are built from the same generalised components.

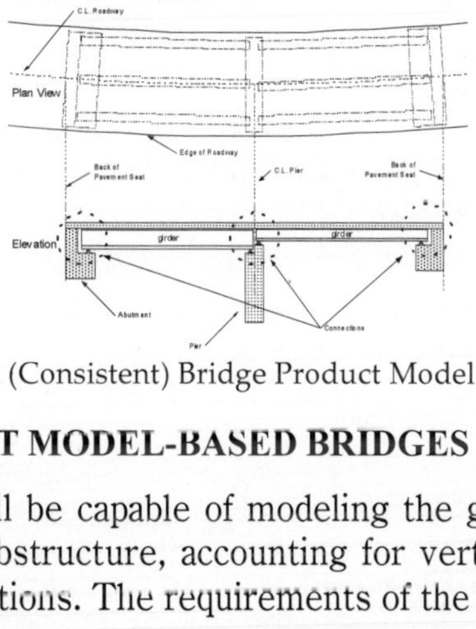

Fig. General (Consistent) Bridge Product Model Description

GENERAL PRODUCT MODEL-BASED BRIDGES

QConBridge II will be capable of modeling the geometry of the bridge superstructure and substructure, accounting for vertical curves, horizontal curves, and superelevations. The requirements of the roadway geometry.

General Requirements

It is impractical to model all possible bridge permutations – we must strive to solve practical problems without undue complications. Therefore, the following general requirements apply to all bridges that can be modeled by QConBridge II:

- When horizontal curves are present, the bridge must be entirely contained within a horizontal curve
- When vertical curves are present, the bridge must be entirely contained within a vertical curve. Bridges can have single or multiple spans.

Roadway Geometry

QConBridge II shall account for the effects of roadway geometry on the structure.

Geometrical considerations shall include horizontal and vertical curves and crown slopes. Effects of the roadway geometry shall include effects on the geometry on bridge components and on the structural behaviour of the bridge.

Roadway Alignment

The roadway alignment is divided into two parts, the horizontal alignment and the vertical alignment. Associated with the alignments is the roadway section that describes the roadway width, crown point offset, and superelevation data.

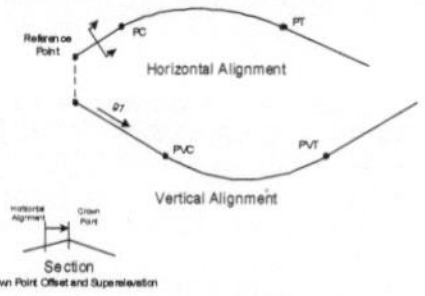

Fig. Schematic of Roadway Alignment

Horizontal Alignment

The horizontal alignment describes the curvilinear path that the roadway follows. The path shall consist of a straight alignment at an arbitrary bearing or a single arbitrarily oriented circular curve.

The reference point for the alignment shall be a known station on a straight alignment or the point of curvature (PC) of a horizontal curve. The horizontal alignment need not represent the centerline of the roadway or bridge. Spiral curve segments are not supported in this version of QConBridge II.

Vertical Alignment

The vertical alignment describes the elevation profile of the horizontal alignment. The vertical alignment is linked to the horizontal alignment at the reference point. The vertical alignment defines the elevation of the reference point and either a straight grade or a vertical curve profile. Compound parabolic curves are not supported in this version of QConBridge II.

Roadway Section

The roadway section is described by its width, crown point offset, and crown slopes. Roadway sections may not change width along the length of the bridge.

Crown Point Offset

The crown point is the high point in the roadway surface. It is located left or right of the horizontal alignment and is constant throughout the alignment.

Crown Slopes

The crown slopes define the slope of the roadway surface. The slopes are measured to the left and the right of the crown point offset. Crown slopes shall be limited to ±0.10.

Superstructure Components

Girderline Geometry

Girders in QConBridge II must be straight in both plan and profile (ignoring camber and haunches).

Adjacent girder lines (if multi-girder) must be parallel. Girders must be aligned in each span such that each girder line is continuous over the entire length of the bridge. Girder centerlines must intersect at the pier centerline.

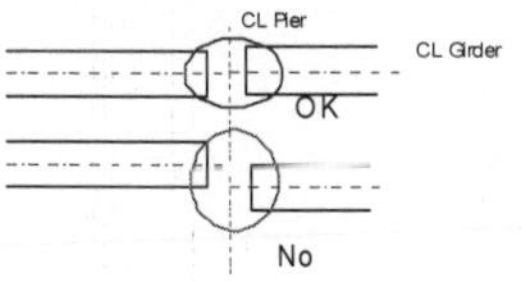

Fig. Continuous versus Discontinuous girder lines

Number of Girders Per Span

The bridge configuration will be limited to bridges with the same number of girders in every span.

Maximum

It is WSDOT practice to label girders A-Z. For this reason, the maximum number of girder lines for all bridge types is 26.

Minimum

A bridge must have at least one girder line.

Girder Spacing

This Section is only applicable for multi-girder bridges.

Minimum Girder Spacing

The minimum girder spacing shall be equal to the width of a girder.

Maximum Girder Spacing

There is no upper limit on girder spacing.

Spacing Types

Girder shall be spacing evenly at each pier and abutment.

Measurement of Girder Spacing

Girder spacing is measured at the location and orientation.

Location of Measurement

Girder spacing will be measured at abutments and intermediate piers. At abutments, girder spacing will be measured at the centerline of bearing line. At intermediate piers, girder spacing will be measured at the centerline of the pier.

Direction of Measurement

Girder spacing can be measured normal to the roadway alignment at the location of measurement or along the centerline of bearing or pier at abutments and piers, respectively.

Span Hinges

An In-Span Hinge consists of a single line of hinges offset from, and oriented parallel to, an associated pier.

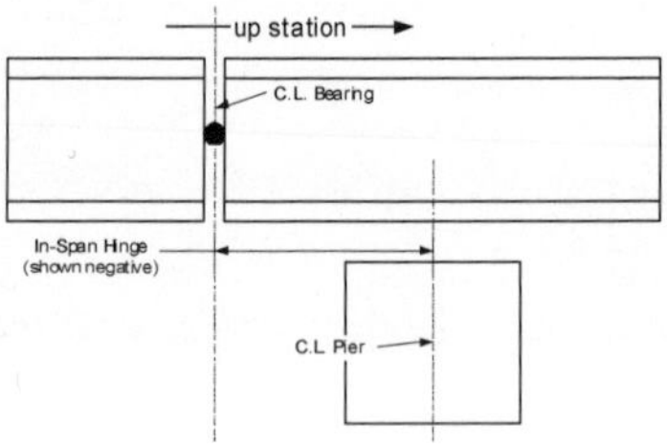

Fig. In-Span Hinge

Span Hinge Geometry

The geometry of an In-Span Hinge is described by the following:

- Hinge Offset The hinge offset is the distance from the pier centerline to the location of the hinge. This distance is measured along the alignment and is positive going up station.

Hinge Boundary Conditions

In-span hinges provide a point of moment discontinuity within a span.

Intermediate Diaphragm Layout Rules

QConBridge II provides two methods for laying out Intermediate Diaphrams: Incremental and Fixed Spacing.

Incremental Layout

Intermediate Diaphragms are spaced evenly along the span length. The number of diaphragms used will depend on the length of the span in question.

Intermediate Diaphragms will be laid out parametrically according to a user-defined set of rules.

Table. Intermediate Diaphragm Layout Rules

Bearing to Bearing Span Length	Number of Intermediate Diaphragms
Less than *L1*	*N1*
L1 to *L2*	N2
L2 to *L3*	N3
L3 to *L4*	N4
Greater than or equal to *L4*	N5

The check boxes indicate optional rules. The starting length of enabled rule N is always the ending length of rule N-1.

FIXED SPACING

Intermediate Diaphragms are laid out using a fixed spacing. After dividing the span into fixed length segments, the remaining distance is divided in half and applied either end of the span.

Substructure Components

One of the driving factors for undertaking the QConBridge II project is the need to compute structural response in substructure elements due to live load. To facilitate LRFD substructure analysis, abutment and pier descriptions must be included in the product model. To facilitate ease of use, QConBridge II will model abutments and piers in a simplified, idealised manner.

Pier types are highly dependent on bridge type. For the purposes of live load analysis, pier models need not be complex. For multi-column piers, simple descriptions of column cross section, column height, column spacing, and crossbeam geometry will suffice. Footings are not modeled in QConBridge II. The point of fixity is assumed to be at the bottom of a column.

Generalised Abutments

In QConBridge II, a general idealisation of an abutment can be made: At the very least, an abutment consists of a line of support crossing the superstructure.

Abutment Location

The station of the back of pavement seat defines the abutment location.

Abutment Orientation

The orientation of the back of pavement seat can be described using three methods: normal to the roadway alignment, or using an absolute bearing value which describes the abutment's direction, or by a Right or Left skew angle measured relative to the normal to the roadway alignment at the back of pavement seat. The absolute value of the skew angle must be less than 90 degrees.

Idealised Abutment Product Models

Idealized product models for abutments provide an expedient way of describing an abutment. There are two types of Idealized Abutments: the Zero-Height Idealized Abutment, and the Fixed-Height Idealized Abutment.

It is very important to note that these are product model descriptions of Abutments, however simplified. It is still up to the Model Generation process to define how these abutment types are represented in the BAM.

Zero-Height Idealized Abutment Product Model

A Zero-Height Idealized Abutment Product Model simply provides a line in space on which to attach a connection. It is representative of a line of support fixed in space at the back of pavement seat. A Zero-Height Idealized Abutment is described only by its location and orientation.

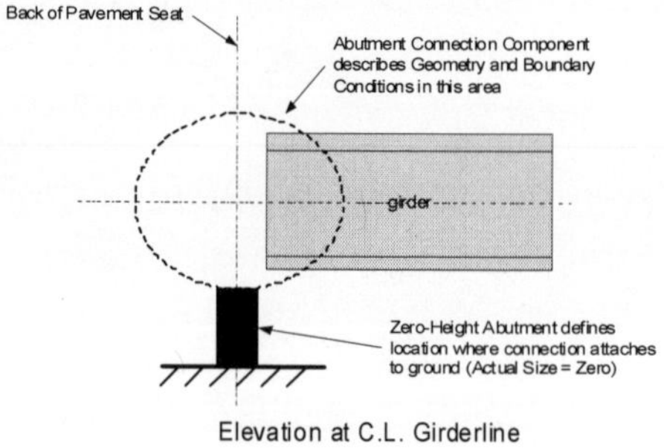

Fig. Zero-Height Idealized Abutment

Fixed-Height Idealized Abutment Product Model

The Fixed-Height Idealized Abutment allows you to define the height and stiffness of the abutment structure into the product model without having to describe a real abutment in detail. Fixed-Height Idealized Abutments are defined by:

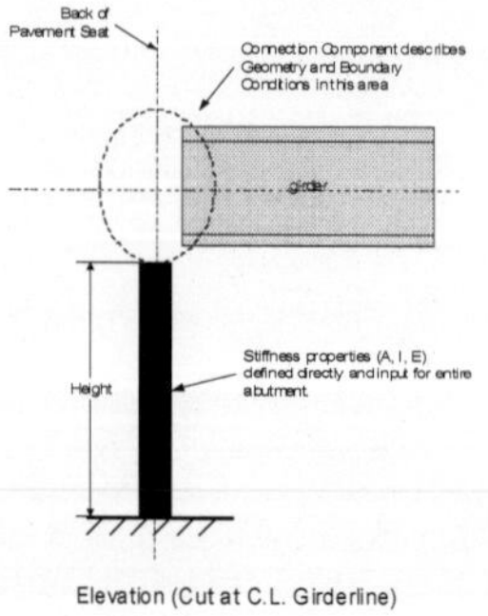

Fig. Fixed Height Idealized Abutment

- Location and Orientation
- Height (measured vertically from bottom of connection to ground) or bottom elevation.
- Stiffness Properties (A, I, E) for entire abutment structure. The model generator will divide stiffness by the number of girder lines to determine the abutment stiffness for a single girder line model.

Figure shows a schematic of a typical Fixed-Height Idealized Abutment.

Piers

QConBridge II will support three types of pier product models. Each type of product module is increasingly more detailed and leads to more complete substructure analysis results.

Zero-height idealized piers provide a line of support to which connections can be attached. Fixed height idealized piers also provide a line of support to which connections can be attached, and include a description of the height and stiffness of the pier. Product Model-Based piers describe a pier in terms of column and crossbeam geometry and materials.

SILO/ TANK CONSTRUCTION

Tanks and silos represent a wide variety of containment structures, most commonly used for the storage of water, grains, bulk-solids, gases or industrial liquids. From engineering design to the supply and installation of stressing and wall-formwork systems, Structural Systems has a strong and successful record in silo tank construction storage structures and silo construction.

HEAVY LIFTING

Heavy lifting could be defined as the erection or relocation of any significant structure without the use of craneage. Typically, such works will involve one or more of cable-based lifting, rod-based lifting and hydraulic jacking.

At SSL, all heavy lifting service projects have the same number one priority - safety. Our track record of successful lifts around the world is impeccable and we aim to keep it that way.

Our achievements reflect a most rigorous process of questioning, checking and monitoring every step of the way - from the initial concept, to component design and on-site execution.

There is no real limit to the size of the load that can be lifted. By example, as Structural Systems has successfully relocated a fragile, single-piece structure weighing 2,000 tonnes, we now like to think that "the sky's the limit!"

Heavy lifting services available from SSL include:

- Engineering design or review
- Hydraulic-ram lifting systems
- Threaded-bar lifting systems
- High-tensile strand lifting systems

LOAD HANDLING

Construction procedures involving load handling systems often result in considerable savings as compared to conventional building methods using traditional scaffolds for casting concrete or installing steel structural elements in place.

The application of load handling systems requires full consideration at the structural design stage, well in advance of detailing or construction planning, as selecting or designing the necessary temporary works and choosing the related equipment must be performed as early as possible.

Structural Systems has extensive experience in the field of load handling and can provide all required services for design as well as supply and operation of equipment.

Typical applications of Structural Systems Load Handling include:

- Lifting, Lowering and Shifting of Heavy Loads Precast beams, entire bridges, roofs, floors and other structural elements built on site or at the factory are lifted in place by means of bars, strands or wire cables and hydraulic jacks.
- Rotating Bridges: Structures built with formwork in a convenient position can be rotated after completion into the final position.
- Working Platforms: Typical for large containments, prestressing tendons and equipment are placed on mobile platforms for installation and prestressing operations in any given position.

- Installation of Large Tendons: In the case of long and heavy tendons, special equipment is required for handling and installation.
- Cable-Stayed Falsework: The truss for building the structure is supported by stay cables replacing expensive temporary supports. Lifting equipment for handling heavy loads is attached to the auxillary tower required for anchoring the stay and backstay cables.

7

Engineering Characterisation of Earthquakes Structure

This section presents information on engineering seismology and engineering characterisation of earthquakes. The key references for this module are Bolt (1988), FEMA (2000), and Kramer (1996). The objective is to introduce the reader to

- Sources and effects of earthquakes,
- Basic concepts and terminology,
- Factors that influence earthquake shaking at a site
- Characterisation of earthquakes for structural engineers
- Attenuation relationships
- Seismic hazard analysis
- Generation of earthquake histories

This module provides an overview of hazard characterisation per the 2000 NEHRP Recommended Provisions (FEMA 2000) so that the reader is familiar with the state-of-practice in the United States.

WHAT IS EARTHQUAKE ENGINEERING?

Earthquake construction is an important discipline that contributes to minimizing damage by using pile foundations, base isolation, and other techniques.

Evaluation of soil conditions is also important. Learn more here.

What Is Earthquake Construction?

Earthquake construction ensures that structures resist earthquake shocks by the integration of seismic designs throughout their expected life, in conformity with the building codes applicable in the region.

Earthquakes destabilize buildings either by direct effects of the seismic waves, or indirectly through soil liquefaction and landslides. Most structures fail laterally by an earthquake, meaning the the walls may fall down, or movement of the walls may cause displacement of the roofs, and result in the collapse of the structure. Therefore, to ensure safety of human life and property, earthquake-resistant techniques should be used, including the utilization of proper design and materials.

Soil Conditions

The condition of soil at the site of construction is an important factor, since the state of soil can significantly alter the motions of an earthquake. The condition of the soil should be thoroughly evaluated. Soils that consist of loose sand and gravel possess poor earthquake-resistant characteristics, and should be reinforced. Seismic waves are amplified in soils that are saturated with water, and change the form of soil from a solid to a liquid upon the occurrence of earthquakes. Such soils acquire the characteristics of quicksand and make the ground incapable of supporting a foundation due to cracks and weakening. Deep and firm soils are good since they allow only minor vibrations to be transferred from the foundation to the construction above.

Pile Foundations

Pile foundations are a structural part used for the transfer

of the structure load to the solid ground located at some depth. Piles are extended and thin elements that transmit the load to a lower soil of greater bearing capability, penetrating the shallow soil. Piles can be used in earthquake construction to minimize earthquake effects, especially with soft surface soils that may easily liquefy, by resisting vertical and lateral loads. A structure is raised on piles if the soil is unstable, weak, does not possess sufficient bearing capacity, and the likely settlement is not advisable. The design of the piles should be proper, by binding the pile caps with reinforced concrete slabs that can function in tension as well as compression, so that the foundation may perform as a unit. In addition, the piles should be designed to carry axial, shear, and bending loads that may be occur because of the horizontal movements between the layers in the soil.

Base Isolation

Base isolation techniques are a recent development in the structural design of buildings and bridges in highly seismic regions. They function on the principles of oscillation and damping. Rubber isolation bearings are used that minimize the earthquake damage to the buildings by decoupling the building from the horizontal component of the ground movement. This is achieved by making the bearings rigid in the vertical direction and elastic in the horizontal direction. The earthquake energy is not absorbed by the base isolation techniques but is deflected due to the system. Rubber bearings can be manufactured easily, do not have any moving parts, and are not affected by time or the environment.

MAKES A BUILDING OR STRUCTURE FAIL IN EARTHQUAKES

An Earthquake moves the ground. It can be one sudden movement, but more often it is a series of shock waves at short intervals, like our ripples from the pebble in the pond analogy above. It can move the land up and down, and it can move it from side to side.

All buildings can carry their own weight (or they would fall down anyway by themselves). They can usually carry a bit of snow and a few other floor loads and suspended loads as well, vertically; so even badly built buildings and structures can resist some up-and-down loads. But buildings and structures are not necessarily resistant to side-to-side loads, unless this has been taken into account during the structural engineering design and construction phase with some earthquake proof measures taken into consideration. This weakness would only be found out when the Earthquake strikes, and this is a bad time to find out. It is this side-to-side load which causes the worst damage, often collapsing poor buildings on the first shake. The side-to-side load can be worse if the shocks come in waves, and some bigger buildings can vibrate like a huge tuning fork, each new sway bigger than the last, until failure. This series of waves is more likely to happen where the building is built on deep soft ground, like Mexico City. A taller or shorter building nearby may not oscillate much at the same frequency.

Often more weight has been added to a building or structure at most frequently at greater heights; say another floor and another over that; walls built round open balconies and inside partitions to make more, smaller, rooms; rocks piled on roofs to stop them blowing away; storage inside. This extra weight produces great forces on the structure and helps it collapse. The more weight there is, and the higher this weight is in the building, the stronger the building and its foundations must be to be resistant to side earthquakes; many buildings have not been strengthened when the extra weight was added. Often, any resistance to the sway loading of the building is provided by walls and partitions;

but these are sometimes damaged and weakened in the Main Earthquake. The building or structure is then more vulnerable, and even a weak aftershock, perhaps from a slightly different direction, or at a different frequency, can cause collapse. In a lot of multi storey buildings, the floors and roofs are just resting on the walls, held there by their own weight; and if there is any structural framing it is too often inadequate. This can result in a floor or roof falling off its support and crashing down, crushing anything below.

Often more weight has been added to building or structure at a higher level, for example another floor, extra walls and partitions, extra storage or even rocks piled on roofs to stop them blowing away. Small cracks appear in the concrete. The bonding of the 'stirrups' (the small steel bars which bind the main reinforcement together) to the concrete weakens, the outer concrete crumbles (spalling), the main reinforcing bars can bend outwards away from the column and all strength disappears. This was beautifully demonstrated under the Oakland Freeway, where huge round concrete columns crumbled and crumpled. They have now been reinforced with massive belts around them as a result of an earthquake engineering review and to improve structural dynamics.

In a lot of multi storey buildings the lower floor has more headroom (so taller columns); and it often has more openings (so less walls); and it is usually stood on 'pinned' feet with no continuity. So the ground-to-first floor columns, which carry the biggest loads from the weight and the biggest cumulative sideways loads from the earthquake, are the longest and the least restrained and have the least end fixity. They are often the first to fail. It only takes one to fail for the worst sort of disaster, the pancake collapse so familiar to any one who has seen the results in Armenia, Mexico, Turkey, Iran, Peru, and now Pakistan and Kashmir. Sometimes buildings are built on soft soil; this can turn into quicksand when shaken about, leading to complete slumping of buildings into the soil. Some tall buildings can stay almost intact but fall over in their entirety. The taller the building, the more likely this is to happen, particularly if the building can oscillate at the frequency of the shock waves, and particularly if some liquefaction of soft soil underneath has allowed the building to tilt.

EARTHQUAKE ENGINEERING

Earthquake engineering is the scientific field concerned with protecting society, the natural and the man-made environment from earthquakes by limiting the seismic risk to socio-economically acceptable levels. Traditionally, it has been narrowly defined as the study of the behaviour of structures and geo-structures subject to seismic loading, thus considered as a subset of both structural and geotechnical engineering. However, the tremendous costs experienced in recent earthquakes have led to an expansion of its scope to encompass disciplines from the wider field of civil engineering and from the social sciences, especially sociology, political science, economics and finance.

The main objectives of earthquake engineering are:

- Foresee the potential consequences of strong earthquakes on urban areas and civil infrastructure.
- Design, construct and maintain structures to perform at earthquake exposure up to the expectations and in compliance with building codes.

A properly engineered structure does not necessarily have to be extremely strong or expensive. It has to be properly designed to withstand the seismic effects while sustaining an acceptable level of damage.

Fig. Shake-table crash testing of a regular building model (left) and a base-isolatedbuilding model (right) at UCSD

Seismic loading

Fig. Taipei 101, equipped with a tuned mass damper, is the world's third tallest skyscraper.

Seismic loading means application of an earthquake-generated excitation on a structure (or geo-structure). It happens at contact surfaces of a structure either with the ground, with adjacent structures, or withgravity waves from tsunami.

SEISMIC PERFORMANCE

Earthquake or seismic performance defines a structure's ability to sustain its main functions, such as itssafety and serviceability, *at* and *after* a particular earthquake exposure. A structure is normally considered*safe* if it does not endanger the lives and well-being of those in or around it by partially or completely collapsing. A structure may be considered *serviceable* if it is able to fulfill its operational functions for which it was designed.

Basic concepts of the earthquake engineering, implemented in the major building codes, assume that a building should survive a rare, very severe earthquake by sustaining significant damage but without globally collapsing. On the other hand, it should remain operational for more frequent, but less severe seismic events.

Seismic performance assessment

Engineers need to know the quantified level of the actual or anticipated seismic performance associated with the direct damage to an individual building subject to a specified ground shaking. Such an assessment may be performed either experimentally or analytically.

Experimental assessment

Experimental evaluations are expensive tests that are typically done by placing a (scaled) model of the structure on a shake-table that simulates the earth shaking and observing its behaviour. Such kinds of experiments were first performed more than a century ago.Only recently has it become possible to perform 1:1 scale testing on full structures.

Due to the costly nature of such tests, they tend to be used mainly for understanding the seismic behaviour of structures, validating models and verifying analysis methods. Thus, once properly validated, computational models and numerical procedures tend to carry the major burden for the seismic performance assessment of structures.

Analytical/Numerical Assessment

Fig. Snapshot from shake-table video of a 6-story non-ductile concrete buildingdestructive testing

Seismic performance assessment or seismic structural analysis is a powerful tool of earthquake engineering which utilises detailed modelling of the structure together with methods of structural analysis to gain a better understanding of seismic performance of building and non-building structures.

The technique as a formal concept is a relatively recent development.

In general, seismic structural analysis is based on the methods of structural dynamics.

For decades, the most prominent instrument of seismic analysis has been the earthquake response spectrum method which also contributed to the proposed building code's concept of today.

However, such methods are good only for linear elastic systems, being largely unable to model the structural behaviour when damage (*i.e.*, non-linearity) appears.

Numerical *step-by-step integration* proved to be a more effective method of analysis for multi-degree-of-freedom structural systems with significant non-linearity under a transient process of ground motion excitation.

Basically, numerical analysis is conducted in order to evaluate the seismic performance of buildings.

Performance evaluations are generally carried out by using nonlinear static pushover analysis or Non-linear time-history analysis.

In such analyses, it is essential to achieve accurate non-linear modeling of structural components such as beams, columns, beam-column joints, shear walls etc.

Thus, experimental results play an important role in determining the modeling parameters of individual components, especially those that are subject to significant non-linear deformations.

The individual components are then assembled to create a full non-linear model of the structure. Thus created models are analysed to evaluate the performance of buildings.

The capabilities of the structural analysis software are a major consideration in the above process as they restrict the possible component models, the analysis methods available and, most importantly, the numerical robustness. The latter becomes a major consideration for structures that venture into the non-linear range and approach global or local collapse as the numerical solution becomes increasingly unstable and thus difficult to reach.

There are several commercially available Finite Element Analysis software's such as CSI-SAP2000 and CSI-PERFORM-3D which can be used for the seismic performance evaluation of buildings.

Moreover, there is research-based finite element analysis platforms such as OpenSees, RUAUMOKO and the older DRAIN-2D/3D, several of which are now open source.

RESEARCH FOR EARTHQUAKE ENGINEERING

Fig. Shake-table testing of Friction Pendulum Bearings at EERC

Research for earthquake engineering means both field and analytical investigation or experimentation intended for discovery and scientific explanation of earthquake engineering related facts, revision of conventional concepts in the light of new findings, and practical application of the developed theories.

The National Science Foundation (NSF) is the main United States government agency that supports fundamental research and education in all fields of earthquake engineering. In particular, it focuses on experimental, analytical and computational research on design and performance enhancement of structural systems.

Fig. E-Defence Shake Table

The Earthquake Engineering Research Institute (EERI) is a leader in dissemination of earthquake engineering researchrelated information both in the U.S. and globally.

A definitive list of earthquake engineering research relatedshaking tables around the world may be found in Experimental Facilities for Earthquake Engineering Simulation Worldwide.The most prominent of them is now E-Defence Shake Tablein Japan.

Major earthquake engineering research centers in the United States and worldwide

Major U.S. research programmes

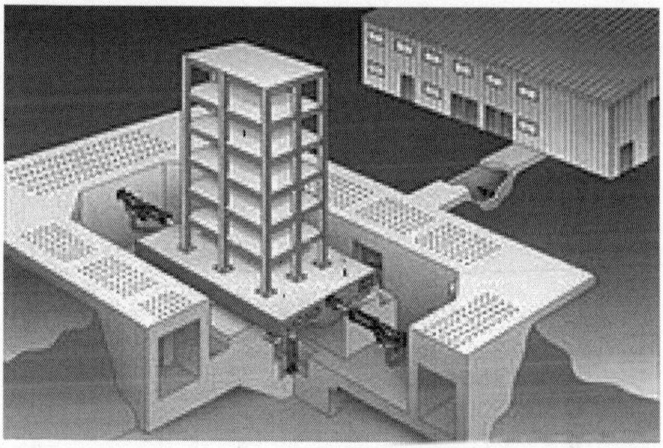

Fig. Large High Performance OutdoorShake Table, UCSD, NEES network

NSF also supports the George E. Brown, Jr. Network for Earthquake Engineering Simulation

The NSF Hazard Mitigation and Structural Engineering programme (HMSE) supports research on new technologies for improving the behaviour and response of structural systems subject to earthquake hazards; fundamental research on safety and reliability of constructed systems; innovative developments in analysisand model based simulation of structural behaviour and response including soil-structure interaction; design concepts that improve structure performance and flexibility; and application of new control techniques for structural systems.

(NEES) that advances knowledge discovery and innovation for earthquakes and tsunami loss reduction of the nation's civil infrastructure and new experimental simulation techniques and instrumentation.

The NEES network features 14 geographically-distributed, shared-use laboratories that support several types of experimental work: geotechnical centrifuge research, shake-table tests, large-scale structural testing, tsunami wave basin experiments, and field site research. Participating universities include: Cornell University; Lehigh University; Oregon State University; Rensselaer Polytechnic Institute; University at Buffalo, State University of New

York; University of California, Berkeley; University of California, Davis; University of California, Los Angeles; University of California, San Diego; University of California, Santa Barbara; University of Illinois, Urbana-Champaign; University of Minnesota; University of Nevada, Reno; and the University of Texas, Austin.

Fig. NEES at Buffalo testing facility

The equipment sites (labs) and a central data repository are connected to the global earthquake engineering community via the NEEShub web site. The NEES web site is powered by HUBzero software developed atPurdue University for nanoHUB specifically to help the scientific community share resources and collaborate. The cyberinfrastructure, connected via Internet2, provides interactive simulation tools, a simulation tool development area, a curated central data repository, animated presentations, user support, telepresence, mechanism for uploading and sharing resources, and statistics about users and usage patterns.

This cyberinfrastructure allows researchers to: securely store, organise and share data within a standardised framework in a central location; remotely observe and participate in experiments through the use of synchronised real-time data and video; collaborate with colleagues to facilitate the planning, performance, analysis, and publication of research experiments; and conduct computational and hybrid simulations that may combine the results of multiple distributed experiments and link physical experiments with computer simulations to enable the investigation of overall system performance.

These resources jointly provide the means for collaboration and discovery to improve the seismic design and performance of civil and mechanical infrastructure systems.

Earthquake simulation

The very first earthquake simulations were performed by statically applying some *horizontal inertia forces* based on scaled peak ground accelerations to a mathematical model of a building. With the further development of computational technologies, static approaches began to give way to dynamic ones.

Dynamic experiments on building and non-building structures may be physical, like shake-table testing, or virtual ones. In both cases, to verify a structure's expected seismic performance, some researchers prefer to deal with so called "real time-histories" though the last cannot be "real" for a hypothetical earthquake specified by either a building code or by some particular research requirements. Therefore, there is a strong incentive to engage an earthquake simulation which is the seismic input that possesses only essential features of a real event.

Sometimes earthquake simulation is understood as a re-creation of local effects of a strong earth shaking.

Structure simulation

Fig. Concurrent experiments with two building models which are *kinematically equivalent* to a real prototype.

Theoretical or experimental evaluation of anticipated seismic performance mostly requires a structure simulation which is based on the concept of structural likeness or similarity. Similarity is some degree ofanalogy or resemblance between two or more objects. The notion of similarity rests either on exact or approximate repetitions of patterns in the compared items.

In general, a building model is said to have similarity with the real object if the two share *geometric similarity*,*kinematic similarity* and *dynamic similarity*. The most vivid and effective type of similarity is the *kinematic*one. *Kinematic similarity* exists when the paths and velocities of moving particles of a model and its prototype are similar.

The ultimate level of *kinematic similarity* is *kinematic equivalence* when, in the case of earthquake engineering, time-histories of each story lateral displacements of the model and its prototype would be the same.

EARTHQUAKE ENGINEERING STRUCTURES

Earthquake engineering structures are such type of engineered structures that withstand various types of hazardous that led to earthquake exposures at the sites of their particular location. Initially, it helps to understand how buildings can be made earthquake resistant by employing the right earthquake engineering practices and considering structural dynamics.

Apart from this, the typical weaknesses and structural deficiencies exists in many buildings that makes more prone to damage and collapse.

Buildings with no Structural Framing

Earthquake damage that are caused by the building having no structural framing exists where the upper floors and roof are simply built on to masonry walls. This is difficult, as some sort of framing is vital; once these walls shake a bit, the entire strength is lost and the building will collapse or pancake during an earthquake. The solution here can only be to strengthen each room with a 12-member cubic frame (on 4 sides round the floor, up the 4 corners of the walls, and around the 4 sides of the roof). Such frames would have to have substantial moment resistance all three ways in all 8 corners. And every room would need the treatment, especially on the lower floors. The frames should be tied in to walls, ceilings or roofs, floors as well as possible, with through bolts, or chemical anchors etc. Such frames should be in steel or in properly designed reinforced concrete. It would perhaps be cheaper to demolish and rebuild properly.

Buildings having Suspect Structural Framing

The buildings have structural framing which is suspect, leading to poor earthquake resistance. This is as common as paragraph 1. You only have to look at pictures from any earthquake to see concrete framing which has failed:

not enough steel main reinforcement, leading to bending failure: not enough stirrups, leading to bursting failure, not enough cement in the concrete, leading to crumbling under load; and of course under-sized members. It is common for people to build hollow block walls, and leave some vertical gaps filled with a bit of concrete with a few rebars; and at every floor, a course of blocks is left out and a similar sized concrete beam replaces it. Unless a structural engineer is quite certain that the concrete and rebars are sufficiently sized and well enough built, especially at the joints, such structures should be treated as in paragraph 1.

Buildings with Floors Resting on Framing

Floors simply resting on framing resulting in complete structural failure during an earthquake. This is very common, even when a proper concrete frame has been used. The floor may be timber joists resting on the walls or cross beams; or resting in joist hangars which have a nail or equivalent holding them in place. Even in reasonable quality flooring, precast concrete planks may be resting on bearing surfaces on walls or beams. Or Steel joists may be simply hooked onto supports on walls or framing with nominal pins. The result is obvious: any slight shaking and the floor falls down. In addition to this, any floor below would not be able to resist the weight of a falling floor; and very often the floors give stability to the framing.

Just look at a video of the World Trade Towers to see how well this works out. The best solution to such problems is to tie adjacent floors together very firmly so that they cannot separate either side of the supports. In a proper frame, these joists are bolted with full moment resisting connections through the main supporting beams; it is not possible to do this in retrospect, but joining adjacent joists over the supports with sturdy bolted steel plates will help: as will breaking out the holes in precast hollow-core concrete planks for a distance of say 400mm either side of the support, and well grouting in deformed rebars across the gap.

Where the joists are supported on outside beams, then similar measure must be adopted to make it impossible for them to fall off.

In almost all buildings, even those designed for earthquake resistance, the shear within the column to beam connections is greater than the shear resistance. This can lead to catastrophic failure.

Excessive Building Weight, Compromising Structural Strength

Additional weight added to buildings leading to structural weakness and elevated susceptibility to earthquake damage or collapse. This is very common, particularly in poor areas of poor countries in earthquake zones. Peru and Haiti spring to mind. A building may have been perfectly well built for one or perhaps two floors. A new roof is put on, usually as described in paragraph 3. Further increases of more floors can follow. If the building does not fail under the continued upwards expansion, it will certainly fail under a slight tremor; even taking down the perhaps well built original building below. In addition, many such buildings have balconies, or have had balconies bolted on; and then the balconies have been bricked in to create further rooms. But the extra weight is in the wrong place. The higher weight is in the building, the more it shakes the building about in an earthquake. Discipline is needed to remove additional weight. If there are block work and concrete roofs, they should be replace with lightweight steel and insulation and sheeting, which may weigh ten times less.

Big Apertures Reducing Building Strength

Big apertures at the lowest floor level compromising the structural integrity of the building particularly during an earthquake. Sometimes, the ground floor level has big apertures in the walls: garage doors, showroom windows, big entrances, internal walls cut away to make impressive lobbies. Often the wall panels that are removed constitute the earthquake sway bracing of the building. Often, inconvenient cross bracing in the new cut apertures is simply cut off, or sometimes replaced with ineffective bracing systems. Often, the ground floor is the highest floor in the whole building, but the columns are a similar construction in every floor. Frequently the columns or walls go down to 'pinned bases'. And any failure of the ground to first floor is likely to be disastrous, and is certainly the worst place to suffer a failure; and everything is conspiring to make this floor also the weakest. To implement earthquake resistance, buildings need to be inspected to ensure that such weaknesses are reinforced, with the re-introduction of bracing. This may best be a four sided square frame of steel or properly reinforced concrete, with fully rigid joints at each corner, vertically in the aperture to give back sway resistance on every wall face with large holes. In all probability the columns from ground floor to first floor also need reinforcing, by clamping steel sections on to them, or strapping round anti bursting concrete around them.

Compromised Building Foundations

The foundations of earthquake resistant buildings require special considerations to allow for ground movement. Frequently the foundations of traditional buildings are often designed as 'pinned', but, because the bottom of every wall or column bears down onto the foundations, the bases do provide a bit of fixity in the static condition. This gives some additional strength to the walls or columns, in the static condition. But in earthquake conditions, the ground is not static: it moves up and down, side to side, and can change slope. If you can imagine a base of a wall or column rotating out of the horizontal, you can see it putting a bend into the wall or column. This means that the apparent fixity at the base is now not giving extra strength but, on the contrary, is contributing to early failure. Also, in earthquakes, the ground can crack and expand or ruck up within the dimensions of a building, and this would put enormous forces into the structure.

For this reason, foundations of buildings in earthquake areas should always have a grillage of reinforced concrete or steel, going both ways under all load supporting members. Such foundations should have full strength connections to the columns, and should be strong enough to give positional and rotational restraint to all the columns. It is not possible to make buildings 'earthquake proof', to the extent that they will resist any earthquake; but the remedies proposed in this list will help any building, and its occupants, survive.

FAÇADE ENGINEERING

Building facades make a major contribution to the overall aesthetic and technical performance of a building. Facade engineers operate within technical divisions of facade manufacturing companies while some engineers provide facade consultancy for:

- architects
- building owners
- cladding
- manufacturers
- construction managers

Projects can include new buildings and recladding of existing buildings.

Facade engineers work in conjunction with other members of the design team including the architect, structural engineer, fire engineer, acoustician and lighting engineers. Their goal is to balance the overall objectives of appearance, quality, performance, cost, durability and schedule. The facade engineer must consider the performance of the facade design with regards to air-tightness, thermal performance that includes heat losses and solar gains, daylight penetration, Acoustic performance, and Fire resistance as well as the overall desired aesthetic for the facade.

Facade engineering requires a blend of skills ranging from structural engineering through to building physics, architecture, manufacturing, materials science, dynamics, programming, procurement, project management and construction techniques.

ROLE OF ENGINEER

Façade engineers come from a range of backgrounds but most usually architectural, structural or building physics. In order then to become a facade engineer, they have then developed a wider breadth of cladding skills and a deeper knowledge than they would encounter within their original discipline. Many façade engineers will be generalist façade engineers. These are able to advise across the full range of materials, systems and performance types.

Specialist façade engineers will typically first have attained a level of knowledge across all façade types and then have chosen to specialise in one particular aspect of façade engineering. Examples are façade engineers whose emphasis is in building envelope physics, using analytical modeling skills; or façade engineers that specialise in a particular cladding material such as stone or glass. Parallels exist with other professions with generalist and specialist divisions *e.g.* legal where there are solicitors and barristers, and medicine where there are general practitioners and consultants. It may be difficult for clients at the inception of their project to decide which type of façade engineer they require. A general practice façade engineer is best placed to determine this for the particular circumstances of a client's individual project and advice on façade specialization that may be needed.

Façade engineers concern themselves with everything to do with a building's external envelope above ground level. Many names can be used to describe the envelope, for instance:

- Cladding
- Curtain wall
- Stonework
- Glass
- Masonry
- Other materials and cladding types

Façade Engineers will consider the performance of such materials and systems in various respects:

- Weather tightness
- Structural behaviour
- Interaction with the primary structure
- Thermal gains and losses through the façade
- Occupant comfort and energy efficiency
- Shading
- Condensation
- Ventilation
- Durability
- Sustainability
- Natural light admittance
- Fire behaviour of the building envelope

- Acoustic performance
- Safety and serviceability
- Security
- Maintenance and buildability

Façade engineers provide advice on both existing and new buildings. They may be involved in design, working alongside the architect, QS and structural and mechanical engineers, or may work within contracting or manufacturing. Alternatively, they may be involved in surveying or diagnostic and remedial work. Some façade engineers are involved in research and testing.

FIRE ENGINEERING

Introduction

Fire which is an integral part of daily living may occur at any moment, anywhere; homes, shops, workshops, motor vehicles, railways, airports or anywhere. Fire when expands becomes dangerous and takes form of disasters, and then services of Fire Safety professionals are called upon.

Fire engineering is the application of science and engineering principles to protect property, people and their environments from the harmful and destructive effects of fire and smoke. It encompasses fire protection engineering which focuses on fire detection, suppression and mitigation and fire safety engineering which focuses on human Behaviour and maintaining a tenable environment for evacuation from a fire. It is observed that the fire protection engineering is often used to include fire safety engineering.

The discipline of fire engineering includes, but is not exclusive to:

- Fire detection - fire alarm systems and brigade call systems
- Active fire protection - fire suppression systems
- Passive fire protection - fire and smoke barriers, space separation
- Smoke control and management
- Escape facilities- Emergency exits, Fire lifts etc.
- Building design, layout, and space planning
- Fire prevention Programmes
- Fire dynamics and fire modeling
- Human Behaviour during fire events
- Risk analysis, including economic factors
- Wildfire Management

Fire protection engineers identify risks and design safeguards that aid in preventing, controlling, and mitigating the effects of fires. Fire engineers assist architects, building owners and developers in evaluating buildings' life safety and property protection goals. Fire engineers are also employed as fire investigators, including such very large-scale cases as the analysis of the collapse of the World Trade Centers. NASA uses fire engineers in its space

Programme to help improve safety. Fire engineers are also employed to provide 3rd party review for performance based fire engineering solutions submitted in support of local building regulation applications.

Fire Engineering is the study discipline, which investigates all possible options of preventing fire and controlling fire in case of a fire breakdown. The reason behind fire breakdown can be any like a loose electric circuit, leakage of a cooking gas cylinder, friction in motor parts etc, exposure of combustible materials to air and so on.

The prevention and control of fire involves use of scientific and engineering know-how in practice, supported by understanding of the causes and effects of fire and the reaction and Behaviour of people, property and the environment, and the impact of fire protection system including detection, alarm and sprinkler system.

Fire engineering profession demands commitments to public services leaving aside the personal safety. The profession is very much lucrative, well paying but involves huge personal risk too.

The role of a Fire Engineer includes engineering design for fire safety and control, operations and management. Fire Engineer work in shifts in highly dangerous conditions and always at risk of exposure to chemicals, burns and smoke inhalation.

They need to be disciplined, quick in action, reliable, leveraged with confidence and full of energy.

At the same time they also need to know about fire fighting, different types of fires and how to prevent and control them, and about rescue methods and equipments in case of a fire breakdown.

The principle responsibility of a fire engineer involves reducing the impact of fire in case of a fire breakdown. Fire Engineers are responsible to improve fire fighting techniques and equipments.

Fire Engineering is the application of scientific and engineering principles, rules and expert Judgement, based on an understanding of the phenomena and effects of fire and of the reaction and behaviour of people to fire, to protect people, property and the environment from the destructive effects of fire. These objectives will be achieved by a variety of means that includes:

- Assessment of the hazards and risks of fire and its effects;
- Mitigation of potential fire damage by proper design, construction, arrangement, and use of buildings, materials, structures, industrial processes, transportation systems and similar;
- Appropriate level of evaluation for the optimum preventive and protective measures necessary to limit the consequences of fire;
- Design, installation, maintenance and/or development of fire detection, fire suppression, fire control and fire related communication systems and equipment;

- Direction and control of appropriate equipment and manpower in the strategy and function of fire fighting and rescue operations;
- Post-fire investigation and analysis, evaluation and feedback.

History

Fire engineering's roots date back to Ancient Rome, when the Emperor Nero ordered the city to be rebuilt utilizing passive fire protection methods, such as space separation and non-combustible building materials, after a catastrophic fire. The discipline of fire engineering emerged in the early 20th century as a distinct discipline, separate from civil, mechanical and chemical engineering, in response to new fire problems posed by the Industrial Revolution. Fire protection engineers of this era concerned themselves with devising methods to protect large factories, particularly spinning mills and other manufacturing properties. One more motivation to organize the discipline, define practices and conduct research to support innovations was in response to the catastrophic conflagrations and mass urban fires that swept many major cities during the latter half of the 19th century. The insurance industry also helped promote advancements in the fire engineering profession and the development of fire protection systems and equipment.

As the 20th century emerged, several catastrophic fires results in change to buildings codes to better protect people and property from fire. It was only in the latter half of the 20th Century that fire protection engineering emerged as a unique engineering profession. The primary reason for this emergence was the development of the body of knowledge, specific to the profession that occurred after 1950. Other factors contributing to the growth of the profession include the start of Institution of Fire Engineers in 1918, and the Society of Fire Protection Engineers in 1950, the emergence of independent fire protection consulting engineer, and the promulgation of engineering standards for fire protection

FIRE ENGINEER

A fire engineer, by way of education, training and experience: understands:

- the nature and characteristics of fire and the mechanisms of fire
- spread and the control of fire and the associated products of combustion
- understands how fires originate
- spread within and outside buildings/structures
- can be detected, controlled, and/or extinguished
- is able to anticipate the behaviour of materials, structures, machines, apparatus, and processes as related to the protection of life, property and the environment from fire
- has an understanding of the interactions and integration of fire safety systems and all other systems in buildings, industrial structures and

similar facilities is able to make use of all of the above and any other required knowledge to undertake the practice of fire engineering.

ROOF ENGINEERING

A roof is the covering on the uppermost part of a building. A roof protects the building and its contents from the effects of weather and the invasion of animals. Structures that require roofs range from a letter box to a cathedral or stadium, dwellings being the most numerous.

In many countries a roof protects primarily against rain. Sometimes the citizens used their roofs for milling wheat, farming, gardens and extra space. Depending upon the nature of the building, the roof may also protect against heat, sunlight, cold, snow and wind.

Other types of structure, for example, a garden conservatory, might use roofing that protects against cold, wind and rain but admits light. A verandah may be roofed with material that protects against sunlight but admits the other elements.

The characteristics of a roof are dependent upon the purpose of the building that covers the available roofing materials and the local traditions of construction and wider concepts of architectural design and practice and may also be governed by local or national legislation.

Designing

For the reason that the roof protects people and their possessions from climatic elements, the insulating properties of a roof are a consideration in its structure and the choice of roofing material.

Some roofing materials, particularly those of natural fibrous material, such as thatch, have excellent insulating properties. For those that do not, extra insulation is often installed under the outer layer. In developed countries, the majority of dwellings have a ceiling installed under the structural members of the roof.

The purpose of a ceiling is to insulate against heat and cold, noise, dirt and often from the droppings and lice of birds who frequently choose roofs as nesting places.

Concrete tiles can be used as insulation. When installed leaving a space between the tiles and the roof surface, it can reduce heating caused by the sun. Forms of insulation are felt or plastic sheeting, sometimes with a reflective surface, installed directly below the tiles or other material; synthetic foam batting laid above the ceiling and recycled paper products and other such materials that can be inserted or sprayed into roof cavities. So called Cool roofs are becoming increasingly popular, and in some cases are mandated by local codes. Cool roofs are defined as roofs with both high reflectivity and high thermal emittance.

Poorly insulated and ventilated roofing can suffer from problems such as the formation of ice dams around the overhanging eaves in cold weather, causing water from melted snow on upper parts of the roof to penetrate the roofing material. Ice dams occur when heat escapes through the uppermost part of the roof, and the snow at those points melts, refreezing as it drips along the shingles, and collecting in the form of ice at the lower points. This can result in structural damage from stress, including the destruction of gutter and drainage systems.

The elements in the design of a roof are:

- the material
- the construction
- the durability

The material of a roof may range from banana leaves, wheaten straw or seagrass to lamininated glass, copper, aluminium sheeting and precast concrete. In many parts of the world ceramic tiles have been the predominant roofing material for centuries.

The construction of a roof is determined by its method of support and how the underneath space is bridged and whether or not the roof is pitched. The pitch is the angle at which the roof rises from its lowest to highest point. Most US domestic architecture, except in very dry regions, has roofs that are sloped, or pitched. Although modern construction elements such as drainpipes remove the need for pitch, roofs are pitched for reasons of tradition and aesthetics. So, the pitch is partly dependent upon stylistic factors, and partially to do with practicalities. Some types of roofing, for example thatch, require a steep pitch in order to be waterproof and durable. Other types of roofing, for example pantiles, are unstable on a steeply pitched roof but provide excellent weather protection at a relatively low angle. In regions where there is little rain, an almost flat roof with a slight run-off provides adequate protection against an occasional downpour. Drainpipes also remove the need for a sloping roof.

The durability of a roof is a matter of concern because the roof is often the least accessible part of a building for purposes of repair and renewal, while its damage or destruction can have serious effects.

SOURCES AND EFFECTS OF EARTHQUAKES

Earthquakes are generated by natural and man-made sources, including

- Tectonic movement
  - Plate boundaries (*e.g.*, San Andreas fault), termed inter-plate earthquakes
  - Mid continent (*e.g.*, New Madrid fault), termed intraplate earthquakes
  - Mid-ocean ridge
- Volcanoes (*e.g.*, Mammoth Lakes, Sierra Nevada mountain range, California)
- Reservoir induced
- Nuclear explosions (*e.g.*, Nevada Test Site)

Earthquakes can damage the built environment a number of ways, including

- Earthquake shaking
- Fault rupture
- Liquefaction or soil failure
- Tsunami (sea) or seiche (lake)
- Flooding
- Fire

Some examples of earthquake-induced damage are presented below.

Fig. Earthquake shaking (Kobe, Japan, 1995

Fig. Earthquake shaking (Kobe, Japan, 1995)

Fig. Earthquake shaking (Izmit, Turkey, 1999)

Fig. Earthquake shaking (Izmit, Turkey, 1999)

Fig. Ground failure (Izmit, Turkey, 1999)

Fig. Tsunami (Alaska, USA, 1964)

Fig. Flooding (Izmit, Turkey, 1999)

Fig. Fire (Izmit, Turkey, 1999)

Fig. Landslide (Northridge, USA, 1994)

Fig. Liquefaction (Kobe, Japan, 1995)

DESIGNING FOR EARTHQUAKE RESISTANCE

For a structure to be earthquake resistant, it doesn't necessarily have to be extremely strong or extremely expensive. Survivability has a lot more to do with the quality of the construction, specifically joints between various components, than it does with overall strength. An important part of earthquake sustainability is dependent upon the flexibility of the materials used in construction. Concrete, a common material used in construction, is not very earthquake resistant. That's because it is extremely strong under compression, but very weak under tension. Earthquakes cause both compression and tension, creating cracks in the concrete. This is why concrete structures are reinforced with steel rods (re-bar), because steel is strong under tension. Pre-stressing concrete can help make the concrete more resilient to earthquakes, as the constant stress on the concrete structure helps prevent it from coming under tension.

Steel structures, such as steel truss bridges are some of the most earthquake proof structures that exist. Not only is steel strong both under tension and compression, but it is somewhat flexible as well. The elasticity inherent in steel allows the structure to flex and still return to its original shape.

One of the technologies which have been developed to help high-rise structures withstand the forces of earthquakes is the Tuned Mass Damper, also known as a Harmonic Absorber. This mass of this weight is determined by careful calculation of the building's weight and design. The intent of the damper is to work on the resonance frequency of the building, not the weight of the building. Located in the upper floors of the building, the weight is coupled to the building with shock absorbers or springs. As earthquakes and other lateral forces (such as high winds) act upon the building, the weight acts according to the first law of physics, not swaying with the building. This helps to dampen the lateral movement of the building. Taipei 101, the world's second tallest building has one of the largest tuned mass dampers ever installed in a building. The 660 metric ton damper is installed between the 87th and 88th floors and suspended from the 92nd to the 88th floor. At the foundation of most skyscrapers, a number of technologies are employed to control how the base of the building interfaces with the foundation. In its simplest form, base isolation has the building sitting on top of the foundation, but not actually attached to it. As the ground and foundation moves, the building resists movement, attempting to stay in one place, according to Newton's first law.

Base isolation can be coupled with a variety of dampers, which act as giant shock absorbers to help isolate the building from the vibrations happening in the ground. Lead rubber bearings, invented by Bill Robinson from New Zealand in 1974 are the current state-of-the-art in base dampening of buildings. These work under the same principle as smaller rubber shock mounts used for motors and other mechanical devices.

BASIC CONCEPTS IN SEISMOLOGY

ELASTIC REBOUND THEORY

The elastic rebound theory proposes that as two plates move relative to the other along a fault segment, elastic strain energy develops in the rock along the plate boundaries, and that *rupture* occurs once the shear stresses in the rock exceed the shear strength of the rock. Because fault planes are generally highly fractured, substantial strain energy can be stored before rupture. If the shear strength of the plate boundary is known, the length of the fault is known, the rate at which the plates are moving relative to one another (termed the slip rate) is known, the time required to build up sufficient strain energy to produce an earthquake and the probable magnitude of that earthquake can be *estimated*.

The illustration of Bolt is of a road running at right angles to the fault. Immediately following construction of the road, the line (ADB) is straight. After time, the line bends with the left side moving with respect to the right side, with the deformation constrained to a relatively narrow width (10s to 100s of meters).

Once the strength threshold of the interface is reached, the fault ruptures and each side of the fault rebounds, that is, point D moves to D1 on the left-hand-side of the fault and D2 on the right-hand-side of the fault.

The effect of fault rupture on a farm fence following the 1906 San Francisco earthquake.

Faulting

Following Bolt, fault displacements can be classified into one of two types: *strike-slip* and *dip-slip*. Bolt illustrates strike slip and dip-slip (normal and reverse) faulting. Faulting is often a combination of strike-slip and dip-slip.

- *Strike-slip*
  - Faulting that produces only *horizontal* displacements along the strike of the fault
- The direction from north of the line of the plane of the fault at the surface is termed the *strike*.
- The arrows on the strike-slip fault below show left-lateral faulting. •

- To determine whether the fault is *left-lateral* or *right-lateral*, imagine that you are standing on one side of the fault line looking across the fault. If the offset on the other side of the fault line is from right to left, the faulting is *left-lateral*. Vice-versa for the other direction.
- *Dip-slip*
  - Faulting that produces *vertical* displacements along the strike of the fault
- 90° dip is vertical
- Two types of dip-slip faults: *normal* fault and *reverse* fault
- *Normal* fault: when the rock on that side of the fault hanging over fracture (the *hanging wall*) plane slips downward
- *Reverse* fault: when the hanging wall moves upwards over the footwall.

SEISMIC DESIGN

Seismic design is based on authorised engineering procedures, principles and criteria meant to design or retrofit structures subject to earthquake exposure.

Those criteria are only consistent with the contemporary state of the knowledge about earthquake engineering structures. Therefore, a building design which exactly follows seismic code regulations does not guarantee safety against collapse or serious damage.

The price of poor seismic design may be enormous. Nevertheless, seismic design has always been a trial and error process whether it was based on physical laws or on empirical knowledge of the structural performance of different shapes and materials.

To practice seismic design, seismic analysis or seismic evaluation of new and existing civil engineering projects, an engineer should, normally, pass examination on *Seismic Principles* which, in the State of California, include:

- Seismic Data and Seismic Design Criteria
- Seismic Characteristics of Engineered Systems
- Seismic Forces
- Seismic Analysis Procedures
- Seismic Detailing and Construction Quality Control

To build up complex structural systems, seismic design largely uses the same relatively small number of basic structural elements (to say nothing of vibration control devices) as any non-seismic design project.

Normally, according to building codes, structures are designed to "withstand" the largest earthquake of a certain probability that is likely to occur at their location.

This means the loss of life should be minimized by preventing collapse of the buildings. Seismic design is carried out by understanding the possible failure modes of a structure and providing the structure with appropriate strength, stiffness, ductility, and configuration to ensure those modes cannot occur.

SEISMIC DESIGN REQUIREMENTS

Seismic design requirements depend on the type of the structure, locality of the project and its authorities which stipulate applicable seismic design codes and criteria.

For instance, California Department of Transportation's requirements called *The Seismic Design Criteria* (SDC) and aimed at the design of new bridges in California incorporate an innovative seismic performance-based approach.

The most significant feature in the SDC design philosophy is a shift from a *force-based assessment* of seismic demand to a *displacement-based assessment* of demand and capacity.

Thus, the newly adopted displacement approach is based on comparing the *elastic displacement* demand to the *inelastic displacement* capacity of the primary structural components while ensuring a minimum level of inelastic capacity at all potential plastic hinge locations.

In addition to the designed structure itself, seismic design requirements may include a *ground stabilisation* underneath the structure: sometimes, heavily shaken ground breaks up which leads to collapse of the structure sitting upon it. The following topics should be of primary concerns: liquefaction; dynamic lateral earth pressures on retaining walls; seismic slope stability; earthquake-induced settlement.

Nuclear facilities should not jeopardise their safety in case of earthquakes or other hostile external events.

Therefore, their seismic design is based on criteria far more stringent than those applying to non-nuclear facilities.

The Fukushima I nuclear accidents and damage to other nuclear facilities that followed the 2011 Tôhoku earthquake and tsunami have, however, drawn attention to ongoing concerns over Japanese nuclear seismic design standards and caused other many governments to re-evaluate their nuclear programmes.

Doubt has also been expressed over the seismic evaluation and design of certain other plants, including the Fessenheim Nuclear Power Plant in France.

Failure modes

Failure mode is the manner by which an earthquake induced failure is observed. It, generally, describes the way the failure occurs. Though costly and time consuming, learning from each real earthquake failure remains a routine recipe for advancement in *seismic design* methods.

The lack of reinforcement coupled with poor mortar and inadequate roof-to-wall ties can result in substantial damage to an unreinforced masonry building. Severely cracked or leaning walls are some of the most common earthquake damage.

Fig. Typical damage to unreinforced masonry buildings at earthquakes

Also hazardous is the damage that may occur between the walls and roof or floor diaphragms. Separation between the framing and the walls can jeopardise the vertical support of roof and floor systems.

Fig. Soft story collapse due to inadequate shear strength at ground level, Loma Prieta earthquake

Soft story effect. Absence of adequate shear walls on the ground level caused damage to this structure. A close examination of the image reveals that the rough board siding, once covered by a brick veneer, has been completely dismantled from the studwall. Only the rigidity of the floor above combined with the support on the two hidden sides by continuous walls, not penetrated with large doors as on the street sides, is preventing full collapse of the structure.

Fig. Effects of soil liquefaction during the1964 Niigata earthquake

Soil liquefaction. In the cases where the soil consists of loose granular deposited materials with the tendency to develop excessive hydrostatic pore water pressure of sufficient magnitude and compact, liquefaction of those loose saturated deposits may result in non-uniform settlements and tilting of structures. This caused major damage to thousands of buildings in Niigata, Japan during the 1964 earthquake.

Landslide rock fall. A landslide is a geological phenomenon which includes a wide range of ground movement, includingrock falls. Typically, the action of gravity is the primary driving force for a landslide to occur though in this case there was another contributing factor which affected the original slope stability: the landslide required an *earthquake trigger* before being released.

Fig. Effects of pounding against adjacent building, Loma Prieta

Pounding against adjacent building. This is a photograph of the collapsed five-story tower, St. Joseph's Seminary, Los Altos, California which resulted in one fatality.

During Loma Prieta earthquake, the tower pounded against the independently vibrating adjacent building behind. A possibility of pounding depends on both buildings' lateral displacements which should be accurately estimated and accounted for.At Northridge earthquake, the Kaiser Permanente concrete frame office building had joints completely shattered, revealinginadequate confinement steel, which resulted in the second story collapse.

In the transverse direction, composite end shear walls, consisting of two wythes of brick and a layer of shotcretethat carried the lateral load, peeled apart because ofinadequate through-ties and failed.

7-story reinforced concrete buildings on steep slope collapsedue to the following:

- Improper construction site on a foothill.
- Poor detailing of the reinforcement (lack of concrete confinement in the columns and at the beam-column joints, inadequate splice length).
- Seismically weak soft story at the first floor.
- Long cantilevers with heavy dead load.

Fig. hifting from foundation, Whittier

Sliding off foundations effect of a relatively rigid residential building structure during 1987 Whittier Narrows earthquake. The magnitude 5.9 earthquake pounded the Garvey West Apartment building in Monterey Park, California and shifted its superstructure about 10 inches to the east on its foundation.

Fig. Earthquake damage in Pichilemu.

If a superstructure is not mounted on a base isolation system, its shifting on the basement should be prevented.

Fig. Insufficient shear reinforcement let main rebars to buckle, Northridge

Reinforced concrete column burst at Northridge earthquake due to insufficient shear reinforcement mode which allows main reinforcement to buckle outwards. The deck unseated at thehinge and failed in shear. As a result, the La Cienega-Veniceunderpass section of the 10 Freeway collapsed.

Loma Prieta earthquake: side view of reinforced concretesupport-columns failure which triggered the upper deck collapse onto the lower deck of the two-level Cypress viaduct of Interstate Highway 880, Oakland, CA.

Fig. Support-columns and upper deck failure, Loma Prieta earthquake

Retaining wall failure at Loma Prieta earthquake in Santa Cruz Mountains area: prominent northwest-trending extensional cracks up to 12 cm (4.7 in) wide in the concrete spillway to Austrian Dam, the north abutment.

Fig. Lateral spreading mode of ground failure, Loma Prieta

Ground shaking triggered soil liquefaction in a subsurface layer of sand, producing differential lateral and vertical movement in an overlying carapace of unliquified sand and silt.

This mode of ground failure, termed lateral spreading, is a principal cause of liquefaction-related earthquake damage.

Severely damaged building of Agriculture Development Bank of China after 2008 Sichuan earthquake: most of the beams and pier columns are sheared.

Large diagonal cracks in masonry and veneer are due to in-plane loads while abrupt settlement of the right end of the building should be attributed to a landfill which may be hazardous even without any earthquake.

Fig. Beams and pier columns diagonal cracking, 2008 Sichuan earthquake

Fig. Tsunami strikes Ao Nang,

Twofold tsunami impact: sea waves hydraulic pressure andinundation. Thus, the Indian Ocean earthquake of December 26, 2004, with the epicenter off the west coast of Sumatra, Indonesia, triggered a series of devastating tsunamis, killing more than 230,000 people in eleven countries by inundating surrounding coastal communities with huge waves up to 30 meters (100 feet) high.

Bibliography

Aditya Kishore Dash and Mira Das: *Advances in Environmental Sciences and Engineering*, Daya Publishing House, 2015.

Anandanatarajan and P. Ramesh Babu: *Control Systems Engineering*, Scitech Pub, 2010.

Appuu Kuttan K.K.:*Control Engineering*, I.K. International Publishing House, 2009.

Bharat Bhushan Prasad: *Advanced Soil Dynamics and Earthquake Engineering*, PHI Learning.

Chan S.: *Park Contemporary Engineering Economics*, PHI Learning.

Constantine A. Balanis: *Advanced Engineering Electromagnetics*, BSP Books, 2008.

D P Das: *Advanced Engineering Mathematics (3 Vols-Set)*, Cyber Tech Pub, 2003.

Gautam Sarkar: *A Textbook of Software Engineering*, Wisdom Press, 2011.

H.K. Dass: *Advanced Engineering Mathematics*, S. Chand Publisher, 2007.

Harish Parthasarathy: *Advanced Engineering Physics*, Ane Books Pvt. Ltd., 2009.

Hazzan: *Agile Software Engineering*, Springer, 2011.

Hiteshkumar V. Parmar: *Agricultural Drainage Engineering*, Scientific Publishers, 2014.

K. Padmanabhan and S. Ananthi: *A Treatise on Instrumentation Engineering*, I.K. International, 2011.

Khan: *Advances in Mechanical Engineering*, Ane Books.

M. Adithan: *Advances in Manufacturing Engineering and Technology*, New Age International, 2010.

Merle C. Potter, J.L. Goldberg and Edward Aboufadel: *Advanced Engineering Mathematics*, Oxford University Press, 2005.

N.K. Dhamsaniya and M.N. Dabhi: *Agricultural Process Engineering (Numerical Problems)*, Agrotech, 2008.

Prabhakar Sharma and S.N. Mukhopadhyay: *A Textbook of Genetic Engineering*, Wisdom Press, 2012.

R. K. Sharma and T.K.Sharma: *A Textbook of Water Power Engineering*, S. Chand Publisher, 2008.

R. N. Yadava, Anil Goyal, Ramakant Bhardwaj and Sarvesh Kumar Agrawal: *Advanced Engineering Mathematics*, BS Pub, 2014.

R. S. Khurmi and J. K. Gupta: *A Textbook of Thermal Engineering Mechanical Technology*, S. Chand Publisher, 2006.

R.N. Reddy: *Agricultural Process Engineering*, Gene Tech Books, 2010.

S. G. Tarnekar: *A Textbook of Laboratory Course in Electrical Engineering*, S. Chand Publisher, 2006.

S. P. Chandola: *A Textbook of Transportation Engineering*, S. Chand Publisher, 2011.

S. S. Bhavikatti: *A Textbook on Elements of Civil Engineering and Engineering Mechanics*, New Age International, 2011.

S. Sivanagaraju and L. Devi: *Control Systems Engineering*, New Age International, 2010.

S.N. Yadav: *Agricultural Engineering: Fundamentals and Applications*, Biotech Books, 2011.

Santanu Ghosh: *An Introduction to Engineering Aspects of Nuclear Physics*, I.K. International, 2009.

Shagufta: *Agricultural Engineering*, APH Pub, 2012.

V.R.L. Gorty: *Advanced Engineering Mathematics Vol. 1*, Ane Books.

Xin-She Yang: *An Introduction to Computational Engineering with MATLAB*, Viva Books, 2008.

Index